CW00448729

FIRST IMPRESSIONS

OF WOLVERTON

Contemporary accounts of Wolverton
by the Early Victorians

compiled and edited by
Bryan Dunleavy

FIRST IMPRESSIONS OF WOLVERTON
Contemporary Accounts of Wolverton by the Early Victorians

Published by Magic Flute Publications 2013

ISBN 978-1-909054-06-6

Magic Flute Publications is an imprint of
Magic Flute Artworks Limited
231 Swanwick Lane
Southampton SO31 7GT

www.magicfluteartworks.com
www.magicflutepublications.co.uk

A description of this book is available from the British Library

CONTENTS

175 Years 1

Part 1: Wolverton before the Railway 4

The Jeffreys Map of 1770 5

Magna Brittannia 5

From Magna Britannia 6

A Tour of the Grand Junction Canal 11

The Bryant Map of 1825 16

The Pigot Trade Directory for 1830 16

The First Ordnance Survey Map of 1834 18

Part 2: How the Railway Came to Wolverton 21

Report by Richard Creed 21

Letter by Geo, Stephenson & Son 22

The Agreement between the Radcliffe Trust and the London and
Birmingham Railway Company 23

Road and River Diversion 28

The Drawings of John Cooke Bourne 29

 Wolverton Embankment 29

 Wolverton Viaduct 29

A Letter from Phillip Hardwick 32

Part 3: First Experiences on the Railway 37

The Correspondence of the Rev. George Phillimore of Willen 37

Morning Post Thursday 20 September 1838 43

Part 4: The Railway Arrives at Wolverton 47

Opening Day 47

A timetable from 1839 50

The First Railway Station 51

Minutes from Some Board Meetings 52

 London and Birmingham Railway Board Meeting 52

 Locomotive Power Committee 52

 London and Birmingham Railway Board Meeting 53

 Coaching and Police Committee 53

The Second Railway Station 54

The Binns and Clifford Survey 54

The Description of Francis Whishaw 56

Extract from the Journal of John Herapath Editor of the Railway Magazine 60

The Census of 1841 63

The Directors and Chief Officers 65

Part 5: The Travellers 69

Drake's London and Birmingham Railway 69

The London and Birmingham Railway 75

Osborne's London and Birmingham Railway Guide 79

The London and Birmingham Railway Companion 82

Part 6: Life in the New Town 93

Hugh Stowell Brown - extracts from Notes on My Life 93

Chapter VIII I Go To Wolverton. 93

Chapter IX Railways in 1840. 95

Chapter X My Mates at Wolverton. 101

Chapter XI I Become a Teetotaller and Sunday School Teacher 104

Chapter XII I Decide to Become a Minister of Religion 106

Building a Community 108

London and Birmingham Railway Board 108

The Radcliffe Trust 109

Consecration of Wolverton Church 111

A Letter from George Bramwell to the The Times 114

Letter from George Bramwell to Ralph Cartwright 114

Letter from the Rev George Weight to Geo. Bramwell 115

The Expansion of Wolverton 117

The Water Supply 125

Salary Registers 132

Matters of Scandal and Public Concern 136

Mr. Blott the Station Master 136

Miss Prince and Mr Russell 138

The Death of a Three Year-Old Boy 139

Entertainment 142

Part 7: The Historians 148

George Lipscomb 149

James Joseph Sheahan.. 153
Part 8: Trade and Commerce 161
Pigot and Company's 1842 Dircetory 161
Kellys Post Office Directory 1847 172
Part 9: The Visitors 179
Hugh Miller. First Impressions of England and its People. 179
Queen Victoria Visits Wolverton 182
The New Road from Wolverton Station to Stony Stratford 186
Sir Francis Bond Head. Stokers and Pokers. 187
Samuel Salt, Railway and Commercial Information........................... 197
Samuel Sidney. Rides on Railways 199
Rambles on Railways 208
Francesca Marton. Attic and Area 214
Part 10: The Close of the Decade 219
Wolverton Mechanics Institute 219

175 YEARS

It is not too far fetched to say the Wolverton, the railway town, came to North Bucks literally "out of the blue". Nothing about it was expected or even understood at the time. Even though the land had been secured for the line on the eastern edge of the Radcliffe estate, there was no intention to build a maintenance depot on the site until as late as 1837 when it became clear that this site, approximately half way between the two termini and serviced by the canal, would be very suitable. At this time a few additional acres were acquired from the Radcliffe Trust and a workshop designed and constructed. Housing for workers came almost as an afterthought in 1838 and Wolverton did not begin to resemble a town until 1840. Up to that point most workers had to find accommodation in the old village or Stony Stratford or Haversham and Bradwell.

The first decade was probably an exciting one as a new community started to emerge. It was a phenomenon - an industrial town in the hitherto under-populated and entirely agricultural area of North Bucks. Naturally, it generated interest, and travellers who made their stopover half way between London and Birmingham all knew about Wolverton. Some of them wrote about it.

The purpose of this volume is to bring together all (or at least most) of the documents I have been able to find about Wolverton Station in that first decade, and to show how Wolverton was seen by those early Victorians. It will also serve as a resource for those wishing to conduct some research into the early history of the town. This is a mixed bag. There are accounts from travellers and professional writers. There are also reports from engineers and various specialists. the compilation also includes letters, newspaper articles, memoirs, contemporary histories, board minutes, census extracts, maps and engravings. Wolverton attracted a great deal of attention in this decade.

The result is a compendium of source material, gathered into a single volume, that is intended as a reference rather than a narrative.

Steam locomotives that had the power to draw goods wagons and passenger coaches along railways were a genuine phenomenon 175 years ago. This can't be overstated; travel was now faster by degrees and beyond the imagination and human experience up to this point. It did wonders for the economy. Goods could be delivered quickly; the Post office delivery service became faster and cheaper. Shops could offer a greater range of goods and keep their keep their stock up-to-date. The railways facilitated the boom in manufacturing and retail industries.

Wolverton had an important part to play and it quickly became a model for other rural railway towns such as Swindon and Crewe. By the end of the century Wolverton Works employed over 5,000 men and supported a town that was second only to High Wycombe in size in Buckinghamshire.

As the twentieth century dawned Wolverton was the largest carriage manufacturing plant in the country but its future from this point was less brilliant. The railway companies' resources were drained by their contributions to two major wars and the government never put back what it took out. The inter-war years were a struggle. Grouping the railway companies into larger enterprises meant that Wolverton had secondary importance from 1924 and there was further decline after nationalisation in 1948. As I write this today there are approximately 100 skilled employees working in rolling stock maintenance and sad to say even their future is in doubt.

After 175 years Wolverton scarcely has a railway presence at all but it does have a proud and important history. This is worth remembering and worth celebrating and I hope that this volume will capture some of the excitement of that first railway decade which began 175 years ago.

Bryan Dunleavy
September 2013

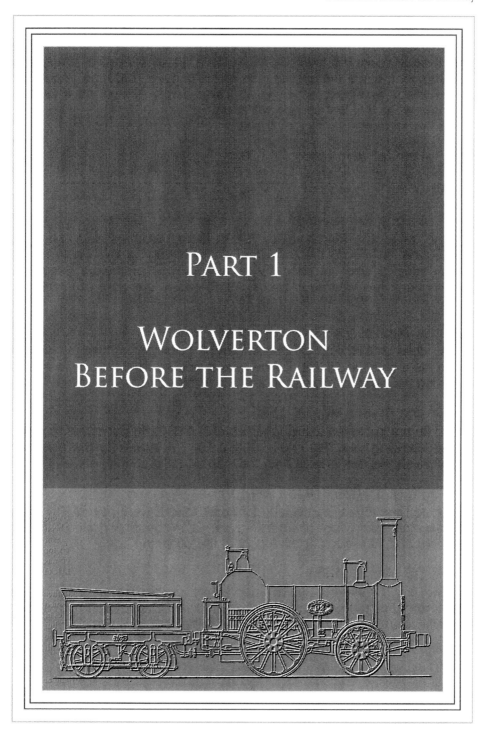

PART 1

WOLVERTON
BEFORE THE RAILWAY

PART 1: WOLVERTON BEFORE THE RAILWAY

Before 1838 Wolverton was a typical 19th century agricultural village with a busy coaching town on its western edge. It was no stranger to development; since the appearance of the canal in 1800 the population of the village had doubled. Even so it was vastly diferent from its later incarnation, as this first section will show.

In this section.

- ## The Jeffreys Map of 1770

Thomas Jeffreys surveyed Buckinghamshire between 1766-8 and published the map in 1770 at a scale of 1 inch to 1 mile

- ## Magna Brittania

The project undertaken by the Lysons brothers was huge in conception, to record the topography and history of the whole of the British Isles. It was cut short by Samuel Lysons premature death but Buckinghamshire was published in 1819, and is therefore of interest to us.

- ## A Tour of the Grand Junction Canal

As does J C Hassall produced several books, illustrated by himself about his travels. in 1819 book he toured the Grand Junction Canal and recorded his views of Wolverton and Stony Stratford.

- ## The Bryant Map of 1825

A. Bryant surveyed Buckinghamshire in 1824 and published the map in the following year. He chose a scale of 1½ inches to the mile and therefore shows more detail than the map of half a century earlier.

- ## Pigot's 1830 Trade Directory

Trade Directories, particularly for small towns, were new at this time. This earliest surviving directory gives us a brief portrait of Stony Stratford and Wolverton at the height of the coaching age.

- ## OS Map of 1834

The first Ordnance Survey map offers us a detailed view of the Wolverton area immediately prior to the railway. This revised map shows the railway and new town as an overlay to the original survey.

The Jeffreys Map of 1770

The map pre-dates the canal and shows much that had been lost even 50 years later. Trackways cross the fields uninterrupted by canals and railway embankments. The river ouse runs its old course. The old manor house at Wolverton has already gone and Wolverton House has yet to make its appearance.The old mansion at Stantonbury is still standing at this time and is given some prominence as the grandest house in the area.

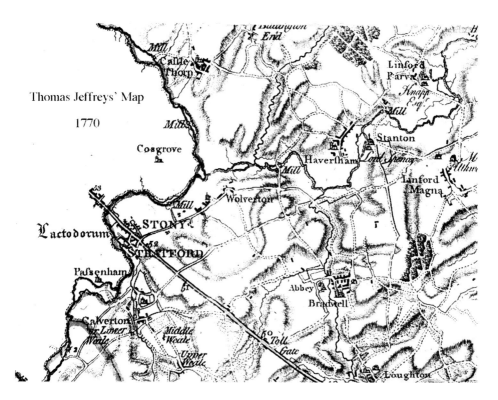

Magna Brittannia

The ambitious account by the Lysons brothers provides us with a view of Wolverton and Stony Stratford in that transition period between the agricultural age and the industrial age.

Daniel and Samuel Lysons were brothers who collaborated on some ambitious books of antiquarian interest. The first major work was *The Environs of London, being an Historical Account of the Towns, Villages and Hamlets within twelve miles of that Capital.* Daniel was probably responsible for most of the text while Samuel contributed engravings. The book was published in several volumes between 1792-6. After its successful reception they embarked on their ambitious *Magna Britannia* which was planned to cover all counties. The project was curtailed due to Samuel's relatively early death in 1819, by which time they had only completed the Counties from B - D.

Fortunately that included Buckinghamshire and the extracts relevant to Wolverton are re-printed here.

The extracts from the book selected here describe the River Ouse, which forms the boundary to North Bucks, the new canal and some words about the roads. Of interest is the reported size of some of the fish caught in the Ouse, including salmon. I doubt if cathces of this size were ever reported in the 20th century. Roads were certainly better than they were a century before, but in the first decade of the 19th century the road from Newport Pagnell to Bedford through Chicheley is described as quite impassable. It is interesting to note also that the road which now borders the east side of Milton Keynes was once considered part of the great road to Liverpool.

At this time too, the whole area is agricultural, except for the prevalence of lace making, a cottage industry that was organised on a large scale in North Bucks at that time.

Wolverton gets a brief paragraph on its ancient history.

From Magna Britannia

The course of the Ouse through Buckinghamshire, or as a boundary to it, is very circuitous, being little less than 50 miles. It first becomes a boundary to this county in the parish of Turweston, near Brackley, separating it from Northamptonshire, and, having passed Westbury, for a few miles divides Buckinghamshire and Oxfordshire: it then enters this county at Water-Stratford, passing near Radcliffe, to Buckingham; from thence to Thornton, a little beyond which it becomes again for a few miles a boundary between Buckinghamshire and Northamptonshire, passing near Beachampton to Stony-Stratford, a little beyond which it enters the county a second time, and passes near Wolverton, Haversham, Stanton-Barry, Linford-Parva, and Lathbury, to Newport-Pagnell; from thence, between Gayhurst and Tyringham, to Olney, leaving Stoke-Goldington, Ravenston, and Weston-Underwood, on the left: from Olney it runs by Clifton-Reynes, and Newton-Blossomville; and, passing between Brayfield and Turvey, forms, for a short distance, a boundary between Bedfordshire and

6

Buckinghamshire, after which it quits this county near Snelson, in. the parish of Lavendon.

We are told that the following fish, of a remarkable size, were recorded on the kitchen walls of the old manor-house, at Tyringham, (now pulled down,) as having been caught in the river Ouse, in this county: a carp, in 1648, measuring 2 feet 9 inches in length; a pike, in 1658, 3 feet 7 inches in length ; a bream 2 feet 3! inches; a salmon, 3 feet 10 inches; a perch, 2 feet; and a (had, in 1683, 1 foot 11 inches1. Salmon, carp, and (had, are not in general reckoned amongst the fish of the Ouse.

The Ousel is a boundary between Bedfordshire and Buckinghamshire, from Eaton-Bray to Linchlade, near Leighton-Busard; where, entering this county, it runs near Stoke-Hamond, and Water-Eaton, to Fenny-Stratford; thence by Simpson, Walton, the two Woolstons, and Willen, to Newport-Pagnell, where it falls into the Ouse. Its course, as connected with this county, (being almost from its source,) is nearly thirty miles. This river is remarkable for fine perch, pike, and bream.

Navigable Canals.

The Grand Junction Canal enters Buckinghamshire near Wolverton, where it is carried across the valley over the river Ouse, which is here the boundary of the county, by an aqueduct of about three quarters of a mile in length: it passes near Linford-Magna, leaving Newport Pagnell on the north; by the Woolstons, Woughton, and Simpson, to Fenny-Stratford; thence leaving Stoke-Hamond, Soulbury and Linchlade, on the. west, it follows the course of the river Ouse to Grove, leaving Leighton-Busard, in Bedfordshire, on the east, and afterwards leaves Cheddington on the west, and Slapton, Ivinghoe, and Marsworth on the east: near the last-mentioned place it quits the county.

In the year 1794 an act of parliament passed for making cuts from the towns of Aylesbury, Buckingham, and Wendover, to communicate with the Grand Junction. Canal.

Roads.

The great road from London to Chester and Holyhead enters Buckinghamshire between the 41st and 42nd mile-stones, (in the course of the ancient Watling-Street,) and passing through Little-Brickhill, Fenny-Stratford, Shenley, and Stony-Stratford, quits the county at Old-Stratford, near the 53rd mile-stone. The great road to Liverpool enters this county near the 43rd mile-stone, about a mile beyond the town of Woburn, and passes through Wavendon, Broughton, Newport-Pagnell, and Lathbury, between Gayhurst and Tyringham, and through Stoke-Goldington, about two miles beyond which it enters Northamptonshire, between the 57th and 58th mile-stones. The road from London to Oxford, Bath, &c. enters this county at Colnbrook, and, passing through Slough, quits it at Maidenhead bridge. The other road to Oxford (commonly called the Wycombe road) enters Buckinghamshire just beyond Uxbridge, and passes through Beaconsfield, High-Wycombe, and West-Wycombe, quitting the county near Stokenchurch, a little beyond the 37th mile-stone. The road from London to Banbury quits the Wycombe road at the 18th mile-stone, and passes through the two Chalfonts, Amersham, the two Missendens, Wendover, Aylesbury, Hardwick, Whitchurch, Winslow, Adstock, Padbury, Buckingham, and Tingewick; a little beyond the last-mentioned place, after passing the 60th mile-stone, it enters Northamptonshire. The ancient course of the road from Aylesbury to Buckingham left Hardwick, Whitchurch, and Winslow on the right; passing through East-Claydon, and between Steeple-Claydon and Padbury, as appears by Ogilby's roads of England and Wales, published in 1736. The turnpike road from Buckingham to Brackley passes by Shalleston and Westbury. The road from London to Aylesbury, through Tring, enters Buckinghamshire, between the 32d and 33d mile-stones, and, passing through Aston-Clinton, joins the other road near Aylesbury. The road from Reading to Ware, in Hertfordshire, enters this county a little beyond Henley, and continues along the banks of the Thames through Medmenham to Great-Marlow; thence to High-Wycombe; over Wycombe heath to Amersham ; and over Amersham common to Cheynies; a little beyond which it enters Hertfordshire. A turnpike road from Aylesbury to Bicester passes through Fleet-Marston, and Waddesdon, quitting the county just beyond Ludgershall common. From Newport-Pagnell a turnpike road passes through Sherrington and Emberton to Olney; and thence, through Cold-Brayfield, to¬wards Bedford. Another road, passing through Weston-Underwood, connects Olney with the Northampton road.

The road from Newport-Pagnell to Bedford, passing through Chicheley and Astwood, which was formerly the principal road from Oxford to Cambridge, and the route of the judges on the midland circuit, is still so described in the Itineraries ; but it has been many years neglected and disused, and is now quite impassable for carriages.

Manufactures.

FULLER, says, that bone lace (an obsolete term for thread lace) was made in the neighbourhood of Olney; but that Buckinghamshire was not to be considered as a manufacturing county, "more people living by the lands than the hands." It is to be presumed, that the lace manufacture has been much extended since Fuller's time , as it is now pretty general in most parts of the county: it is still carried on to a great extent in and about Olney, where veils, and other lace of the finer sort, are made, and great fortunes are said to be acquired by the factors. Lace-making is, in no part of the county, so general as at Hanslape, and in its immediate vicinity; but it prevails for 15 or 20 miles round, in every direction. At Hanslape not fewer than 800, out of a population of 1275, were employed in it, in the year 1801: children are there put to the lace schools, at or soon after five years of age; at eleven or twelve years of age, they are able to maintain themselves without assistance: both boys and girls are taught to make it; and some men, when grown up, follow no other employment: others, when out of work, find it a good resource, and can earn as much as the generality of day-labourers: the lace made at Hanslape is from 6d. to two guineas a yard in value. It is calculated, that from £8,000 to £10,000 neat profit is annually brought into this parish by the lace manufacture.

WOLVERTON, in the hundred and deanery of Newport, lies about a mile north-east of Stony-Stratford : it was the seat of the barony of Maigno Brito, a powerful Norman, whose descendants took the name of Wolverton: the family became extinct in the male line, in the reign of Edward III. John Longueville, who died in 1439, was possessed of this manor by marriage with Joan, daughter and heir of John Hunt, by his wife Margaret, daughter and sole heir of Sir John de Wolverton. The manor of Wolverton continued in the family of Longueville, nearly 300 years. Sir John Longueville, who was owner of Wolverton in Leland's time, lived to the age of 103: his descendant, Sir Edward Longueville, was created a baronet of Nova Scotia in 1638, being described of this place: the title is now extinct. Sir Edward Longueville, the last baronet but one, broke his neck by a fall from his horse at Bicester races in 1718, and it is remarkable, that his father Sir Thomas, met with his death by a similar accident in

1685. About six years before his death, Sir Edward Longueville above mentioned, sold Wolverton to the celebrated physician, Dr. Radcliffe, who bequeathed it with other large estates in trust for the university of Oxford.

The keep of Maigno Brito's castle remains near the vicarage. The seat of the Longuevilles, which was re-built in 1586, has been pulled down: Browne Willis describes it as a magnificent mansion.

In the parish church is the monument of Sir Thomas Longueville above-mentioned, who died in 1685: numerous entries of births and burials of this' family occur in the parish register.

The great tithes which were appropriated to the Priory of Bradwell, were granted by Queen Elizabeth to Sir John Spencer, whose grandson, Spencer Compton, Earl of Northampton, sold them about the year 1650 to the Longuevilles: having been included in Dr. Radcliffe's purchase, and in his bequest to the university, they are now vested in the trustees under his will, who are patrons of the vicarage.

The priory of Bradwell adjoining to this parish, the site of which is now deemed extraparochial, was founded in the reign of King Stephen, for black monks, by Manfelin, Baron of Wolverton: it was dedicated to the Virgin Mary, and was originally a cell to Luffield. In 1526 it was given with other small monasteries to Cardinal Wolsey; after his attainder, the king granted it with the manor, in the year 1531, to the prior and convent of Sheene: the site was: granted after the reformation, to Arthur Longueville esq. From the Longuevilles it passed by purchase to the Lawrences, in 1647; and from them, in 1664, to Sir Joseph Alston bart. then of Chelsea, who made Bradwell Abbey his residence: after his death it was successively in the families of Fuller and Owen. About the year 1730 the Bradwell Abbey estate was purchased by Sir Charles Gunter Nicholl, K. B. whose only daughter and heir married the late Earl of Dartmouth: it is now vested in their son, the present earl. The site of the abbey, of which there are no remains, is occupied as a farm-house.

A Tour of the Grand Junction Canal

In the year of Samuel Lyson's death J C Hassall published his tour of the Grand junction canal. We can get an immediate account of Wolverton prior to the railways from John Hassall, an early 19th century travel writer who made a journey along the new Grand junction canal in 1819 and wrote down and drew his observations. The drawing he left of Wolverton was from a still more-or-less recognizable view point beside the Wolverton turn just after the Galleon bridge. The public house, first called the Locomotive and later called the Galleon had yet to be built but we get some wharfside cottages and a view to Cosgrove Hall in the distance. It is a very bucolic scene, with only a hint of the industralization that came with the canal. But let Hassall decribe what he found.

Passing Stone-bridge-house, by a very pleasant road, shaded with an abundance of lofty trees, we come to the Village of Woolverton. The country here burst upon us with peculiar beauty; the scenery on our right presented the navigation passing in long line away to Cosgrove, which terminates in a succession of woods, towering above each other, with the seat of Major Mansell to the left, and Cosgrove church a little more to the right. The Grand Junction is here carried across the valley by an embankment originally built upon arches, but bursting its course, a wooden trough was afterwards substituted, through which the barges were conveyed to either part of the canal. This trough has given way to one made of cast iron, for the special purpose of uniting the channel, the durability of this metal being much greater than that of wood. The artificial channel is much narrower than the usual passage of the navigation, and viewed from the bridge at this village, has a singular appearance. The component parts of the scenery about Woolverton are truly picturesque, and afford a variety of subjects for the pencil. The Ouse river that runs from Old Stratford passes up the vale before us, and under the channel of the Grand Junction canal, where it unites with the river Tone, which comes from Stoke Bruern park, the seat of Mr. Vernon, and then takes its course by Haversham and Stanton to Newport Pagnel, in a north-east direction.

While sketching the scene before us, a hare which had just been shot at, passed close by the bridge pursued by a pointer; having been headed by another dog, it returned down the hedge, and making two doubles, went into some rushes on the bank of the navigation, from whence it swam across the canal and made for the woods. By this time the sportsmen had got up with their dogs, who instantly pointed at the rushes; at length they encouraged their pointers to go in among them, when all appeared disappointment; and thejr observing me at a short distance, probably imagined poor puss had died of her wounds at that spot, and that I had witnessed some person take her away. In full confidence of their opinion, the gentlemen politely addressed me on the subject, when I satisfied them by pointing out the poor creature, who had again been disturbed, and was coursing the opposite valley in a fresh direction, but at too great a distance for their pursuit.

A view to what was later known as *The Galleon Bridge*, Old Wolverton

According to Camden, this village was anciently called Wolverington, from the seat of an ancient family so named, whose lands are called in the records, the barony of Wolverington ; from them it descended to the Longuevilles, many ages ago; for John de Longueville, who was sheriff of the county (18 Richard II.) had his residence here. It was lately in the possession of Sir Edward Longueville; it was afterwards purchased by the late famous Dr. Ratcliffe, who was so shamefully treated at the death of Queen Anne as to cause his dissolution.

Prior to the present improvement the Grand Junction was passed between Cosgrove and Woolverton by nine locks, which considerably retarded the navigation, and caused the company to contract with some persons at Stony-Stratford for erecting a new embankment across the valley, which was to be made as durable as Warwick-bridge, with similar abutments, the Grand Junction company, reserving to themselves the right of inspecting surveyors and engineers to determine its durability and workmanship; a previous trial of twelve months was also to be allowed the Grand Junction company to prove its solidity. After its erection Mr. Bevan, the engineer, of Leighton Buzzard, being called upon, gave it as his opinion, it would not stand twelve months; his prediction was verified, for in less than six months after its construction, the materials were so indifferent, that a continued leakage of the aqueduct was observable; which occasioned a wag of Stony-Stratford, to observe, "the drainage and droppings of the water were the tears of the contractors of the valley;"— a joke, the parties, it is said, have never forgiven.

Mr. Bevan suggested, after the blowing up of the embankment, the temporary erection of a wooden trough, until another of iron could be cast, which was undertaken and finished at Heseltine's foundry in Shropshire, and laid down in the place of the wooden one, all in the course of twelve months from the time of the accident.

Mr. Cherry, of Greenbridge lock, near Woolverton, was the first person to observe the disaster, and at eleven o'clock at night had but just time to pull up the stop gates, and let off some of the waste water, before the embankment blew up. He sent off a messenger to apprise the inhabitants of Stony-Stratford of the accident. The consternation soon became general, every inhabitant expecting momentarily bis house to be insulated from the effects of the approaching element. The alarm, added to the time of night, caused a dreadful and awful suspense, which only subsided with the day-breaking, when it was observed the valley only was inundated; which cleared off its waters in about three days. Fortunately the graziers lost no cattle of any consequence from its effects.

The devastation occasioned at this spot, by the bursting of the canal, reminds me of a calamity that befel the late Duke of Bridgewater. While prosecuting his favourite undertakings in Lancashire, he had the mortification to witness a breach made by the blowing up of a dam, which instantaneously covered the surrounding country, and it was supposed by all who witnessed the irruption, that it was completely fatal to the whole work. Astounded at the event, he sat motionless on the bank, and petrified with surprise, while viewing the work of years, ruined in an instant. This was a stroke of fate that few even of your boasted philosophers could have withstood—an immense fortune embarked in a speculation, which in an instant appeared blasted beyond all hopes of recovery. His great mind was now strained to its utmost bearing, and he determined at the moment to complete his task or sacrifice all he possessed. He accordingly mortgaged and borrowed money by every possible means, placed hundreds of additional hands to clear away the sand and rubbish which had been heaped together, and in a short period he had the consolation of beholding, finished, what his perseverance and resolution had determined him to follow up. It is still to be hoped a nations' gratitude will cause to be erected an adamantine statue to perpetuate the memory of this great and distinguished nobleman, of whom, it may truly be said, his whole life and fortune were devoted to his country's welfare.

Woolverton church is a very handsome structure, partly of a modern Saxon order of architecture, with curious circular frame-work to hold the glasses. The whole village and its accompanying scenery are beautiful to a degree. A mile from hence we enter Stony-Stratford, which stands upon the old Roman causeway, the Watling street. It is a populous and well frequented market-town, but is most vilely paved with stones of various dimensions. This being the great thoroughfare to Chester, Wales, and Ireland, it is not a little surprising that it should be suffered to remain in such a wretched state, and more so, that the town should not be indicted for neglect of the highway. Camden and Stukeley differ materially about this town, the former averring it to be the Lactodorum of the Romans, and the place where Edward I. erected a cross to the memory of his queen, Eleanor of Spain, adorned with the arms of England, Castile, and Leon; as Hollinshed says, he did in all other towns between this and Westminster, where the corpse rested. While on the other hand, Stukeley affirms, Lactorodum to be Old Stratford, on the opposite side of the Ouse to Stony Stratford, and that Queen Eleanor's cross stood a little north of the Herat Shoe Inn, pulled down in the Rebellion, which, he says, shews the town was on that side of the bridge in the time of Edward I. In and about this neighbourhood, he also observes, there has been found a number of Roman coyns.

The Market Square at Stony Stratford in 1819

Opposite the Cock, the most celebrated inn at Stratford, stands St. Giles's, the largest of the two parish churches; the west end of this building is terminated by a triangular projecting Gothic window of beautiful workmanship ; the inside is a miniature of Westminster Abbey; the pillars that support the roof are remarkably small in their circumference, though very lofty. It is a very tasteful building, and does infinite credit to Mr. Irons, the architect of Warwick. It was built in the years 1776 and 1777; it was formerly a chantry, valued at £20 2s. 6d. per annum. On the 19th of May, 1742, the town suffered greatly by fire, nearly two-thirds of the east side were consumed, together with the body of the church of St. Mary Magdalen, but the tower is yet standing. The necessary regulations to preserve the peace are made by two of the neighbouring magistrates, who hold their meetings on the first Friday in every month. From this town a branch of the Grand Junction passes away to Buckingham.

Stony Stratford carved from the roadside of the two manors of Wolverton and Calverton was a well established coaching town by the early 19th century. Road improvement must have brought some prosperity to the town. It had two industries, the coaching trade and its spin offs, and lacemaking - a more traditional rural craft.

The Bryant Map of 1825

The 1825 map shows the changes made since the introduction of the canal. There is a new road up the hill from the Newport Road to Bradwell, presumably to accommodate the windmill and the wharf. Some of the other roads have change or even disappeared in the intervening half century. The river ouse follows its pre-railway course with a mill race for Mead Mill.

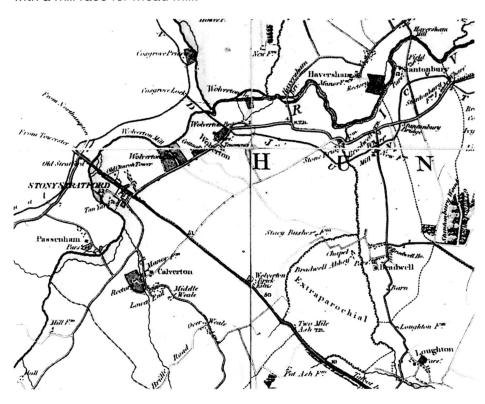

The Pigot Trade Directory for 1830

J. Pigot was an early compiler of Trade directories, a by-product of the expansion of manufacturing industries in the early 19th century. As those markets grew the new breed of travelling salesmen needed information about these new markets. Several publishing companies like Kelly, Pigot, Slater and Baines were pioneers in the field. Kelly outlasted them, possibly because Kelly was astute enough to take over the publishing of the London Post Office Directory in 1835. The earliest trade directory we have in this area is Pigot's directory for 1830. Here he outlines the salient points of the town. Wolverton Manor gets the barest mention, there being no trade or commerce there to speak of.

The coaching trade and lace making were Stony Stratford's staple industries at this date. A decade later, the coaching trade had collapsed after the railway arrived and hand lace making was being replaced by machine made lace in new factories.

STONY STRATFORD, a small market town, and one of some antiquity, is in the hundred of Newport, and the united parishes of St. Mary Magdalen and St. Giles; 51 miles from London, 14 from Northampton, 8 from Buckingham, and 6 from Newport Pagnell. It is situated upon the Watling-street of the Romans, and close to the river Ouse, which here forms the boundary of the counties of Buckingham and Northampton. The parishes are in two manors : the west side being in that of Calverton, the property of William Selby Lowndes, Esq. and the east side in Wolverton manor, belonging to the trustees of the Ratcliff's charity. The government of the town is in the resident and County magistrates, assisted by the officers appointed by the manorial courts. Lace is the manufacture of this place, nearly all the lower class of females being engaged in its production; but its chief trade arises from its thoroughfare situation.

It had once two churches—one dedicated to Saint Giles, the other to Mary Magdalen : the latter fell a prey to a destructive fire, in 1742, which consumed with it 113 houses ; and the church has not been rebuilt since ; the tower, which escaped the fury of the flames, is still standing. The living of the united parishes is a perpetual curacy, in the presentation of the Bishop of Lincoln; the present incumbent is the Rev. Charles Kepling. There are, besides the church, three dissenting chapels, for the independent and Wesleyan methodists, and baptists. A good national school affords instruction to the children of the poor, and there is a charity of £70. per annum appropriated to the apprenticing children, besides several other benevolent bequests, from which the poor of this town receive occasional assistance. It was in this town that King; Richard III, then Duke of Gloucester, accompanied by the Duke of Buckingham, seized the person of the young king, Edward Vth, who was then with his attendants at an inn here. The ground around the town is rather hilly, and the views from some of the eminences are pleasing; the soil is very fertile. The market-day is Friday, and there are two fairs annually, viz. August 2nd, for toys, hardware and pleasure; and the first Friday after Michaelmas-day, for hiring servants. By the population returns for 1821, the number of inhabitants was : east side, 530; west side, 969: total in the united parishes, 1,499.

OLD STRATFORD (Northamptonshire) is a small village, divided from Stony Stratford by the river Ouse. A branch of the Grand Junction canal passes through here, which is of some importance to the agricultural trade of Stony Stratford and the neighbourhood. Neither church or chapel are here, nor is there any appearance of trade. The population returns it is conjectured are made up with those of Stony Stratford.

The First Ordnance Survey Map of 1834

This later edition of the survey first undertaken in 1834 shows the new railway line and New Wolverton overprinted. The area is now at the beginning of its transitional phase, which will be described in the next section.

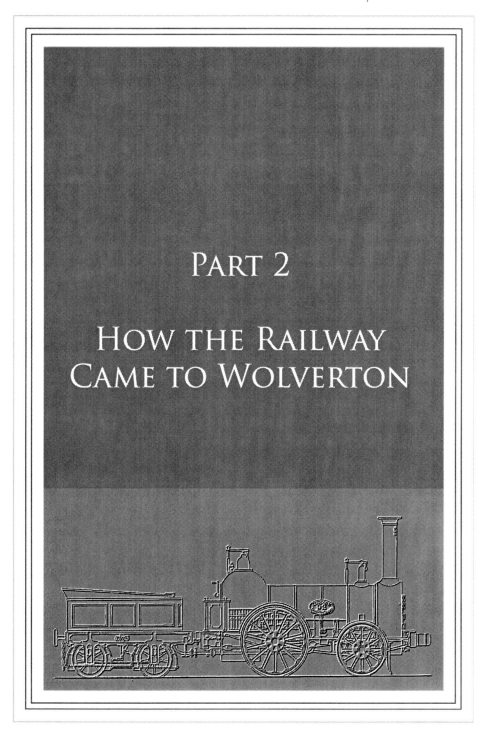

PART 2

HOW THE RAILWAY
CAME TO WOLVERTON

In this section

- A report from Richard Creed to the London and Birmingham Railway Board, dated July 4th 1831

Early exploration of a route through Buckingham.

- A letter from Robert Stephenson to the London and Birmingham Railway Board, dated September 21st 1831

Stephenson's assessment of the terrain via Leighton Buzzard and Rugby along the line that was eventually adopted.

- The Agreement between the Radcliffe Trust and the London and Birmingham Railway Board for the sale of land for the Wolverton depot, dated 3rd May 1837

This agreement, transcribed here, was the first legal document that established Wolverton as a maintenance depot.

- Road and River Diversion

The viaduct across the River Ouse required diversion of the course of the river and a deviation of the road to Haversham.

- The Drawings of John Cooke Bourne

Bourne's beautiful drawings of the construction embankment and the completed viaduct are reproduced in this section.

- A Letter from Phillip Hardwick

The Company architect was asked to report on the progress at Wolverton. This is a snapshot of the state of things at Wolverton only three months before the line was fully open.

PART 2: HOW THE RAILWAY CAME TO WOLVERTON

In the years preparing for the monumental enterprise of a railway linking London and Birmingham the choice of route was by no means clear cut. Here are two reports, one from Richard Creed, the Company secretary, and another from Robert Stephenson considering the implications of a line to the west of Buckingham or one to the east via Leighton Buzzard. In the end, opposition from the Duke of Buckingham scotched the western route and the Stephenson line was adopted, crossing the eastern edge of the Wolverton Manor.

Report by Richard Creed
to the London and Birmingham Railway Board July 4th, 1831 (extract)

Looking towards Birmingham from Ivinghoe here is a breadth of land lying between Leighton Bussard on the East, and Aylesbury on the West, which extends northwesterly to the Ouse and presents formidable obstacles to a Railway. In the direction, therefore, of one or other of these towns it should be carried, and certainly the fertile vale of Aylesbury with its wide spread expanse of level ground, affords great inducement to look through it - Mr. Stephenson having besides called my attention to this point, I proceeded from Ivinghoe, in the first place, towards Aylesbury, and advancing thence on the road to Buckingham, by the high grounds of Whitchurch, Wing and Winslow. I extended my observations sufficiently to the westward to ascertain where the levels most favourable for our purpose were to be found.

Having examined the ground thoroughly on both banks of the Ouse west of Buckingham, and satisfied myself as to the direction the line should take, I followed it to the town of Brackley, and thence along the stream as far up the church of Farthinghoe, where I found the ground to be considerably higher than the summit level of Messrs. G. & R. Stephenson's line, with a rapid descent on the Northwestern face of the range of heights of which this formed a part. Quitting the direction of the line at this point for the double purpose of ascertaining the character of the country about Banbury, and obtaining a base for further observations at the Oxford Canal, I resumed the examination of the same range by Chacombe, Thorpe, Mandeville, the Magpie, and Sulgrave, proceeding thence by

Wormleighton to the Oxford Canal, of which the level afforded the means of correcting the observations made in the course of the day, and by Ladbroke to Southam.

Letter by Geo, Stephenson & Son
sent from Leighton Buzzard, 21st. September 1831

The surface of the country from London and Birmingham may be regarded as consisting of a series of basins or low districts, separated from each other by considerable ridges of Hills.

Having satisfied ourselves that this was the true nature of the surface our next object was to cross the low districts at as high a level and the Ridges of Hills at as low a level as was consistent with the other considerations that governed us with Directness and Cheapness.

The low districts or basins that we have just alluded to are, the London basin from which the line commences. The Valley of the Colne extending from Brentford by Watford to St. Albans. The low land in the neighbourhood of Leighton Buzzard which are connected with districts of a similar level in the direction of Fenny Stratford, to Stony Stratford thence to Stoke Bruern whence it terminates, and is succeeded by the valley of the Line in which Northampton is situated. - The only remaining basin of consequence is that of the Avon; which from its low level - great Breadth - and abrupt terminations (on the South of the high ridge of hills upon which Daventry, Kilsby and Crick are situated and on the north side by the Meriden Ridge) required particular attention.

The high ground or range of hills which form the boundaries of the low districts just described are, the County Boundary between the London basin and the Valley of the Colne - The Chalk ridge at Ivinghoe which rises between the latter and the leighton Buzzard district - The Stoke Bruern and Blisworth which forms one side of the Valley of the Line - The Kilsby and Meriden ridges forming the abrupt sides of the valley of the Avon on the North and South.

The only situation where Tunnelling will be required to any extent is near Cassiobury and through the high ground between Kilsby and Crick. - In other situations where Tunnelling is likely to be adopted, they will scarcely exceed 300 yards in length when the Character of a Tunnel almost vanishes.

The Agreement between the Radcliffe Trust and the London and Birmingham Railway Company

The agreement dated 3rd May 1837 between the Radcliffe Trustees and the directors of the London and Birmingham Railway Company represents the first step in the establishment of the new town of Wolverton. The company had already purchased the land for the line but at this date they sought more land for a maintenance depot, although the word station is used throughout the legal document. They acquired about 8 acres for the large sum of £1,600.

The Radcliffe Trust is keen to assert its ancient manorial rights and insist that any residential dwellings erected on the site should be for railway company employees only. Interestingly, there is a clause insisting that no inn, tavern or be permitted on railway property. The railway Company complied with this and the inevitable taverns that quicly spang up, The Radcliffe Arms and The Royal Engineer, were both built on Trust land outside the railway property.

The agreement, deposited in the Bodleian Library, is a bit damaged, but mostly legible. I have provided a transcription.

While we tend to count the first day the railway line fully opened, 17th September 1838) as the beginning of railway Wolverton, this is really the founding document.

It seems apparent that the decision to establish a depot at Wolverton originated in October 1836, when Edward Bury and Robert Stephenson were out scouting the line for suitable sites for stations. The presence of the canal at this approximate mid-point on the line was probably the clincher with the added advantage that Stony Stratford was only two miles away. Both men recognised not only that a repair depot was required for this infant technology but also that engine crews would need to rest after a few hours on an open footplate in all weathers. The concept of Wolverton was put to the Board and negotiations began with the Trust for the purchase of Roger's Holm.

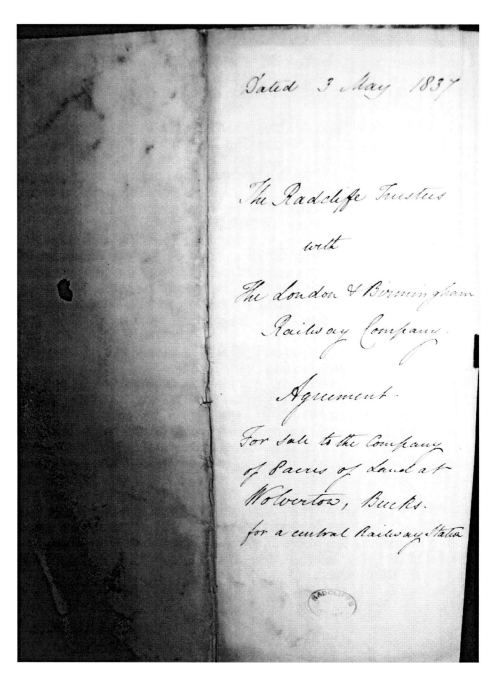

Dated 3 May 1837

The Radcliffe Trustees

with

The London & Birmingham
Railway Company.

Agreement.

For Sale to the Company
of 8 acres of Land at
Wolverton, Bucks.
for a central Railway Station

The Agreement for sale of land in 1837 which began a new chapter for Wolverton

24

A transcription of the Agreement

ARTICLES OF AGREEMENT made this third day of May in the year one thousand eight hundred and thirty seven Between the Right Honourable Sir Robert Peel Baronet William Ralph Cartwright Esquire William Henry Ashurst Esquire and Thomas Gainston Bucknall Escourt the present Trustees of the will of John Radcliffe of the one part and The London and Birmingham Railway Company, stablished and incorporated by an act of Parliament passed in the third year of the reign of His present Majesty and entitled "An Act for making a Railway from London to Birmingham" of the other part **Whereby** it is agreed that this said Trust shall sell and that the said Company shall purchase and for the price or sum of one thousand six hundred pounds two pieces of land belonging to the said Trustees in the Parish of Wolverton in the County of Buckingham situate on each side of the line of the said Railway the one piece containing five acres one wood and twenty seven perches (little more or less) and the other piece containing two acres three roods (little more or less) severally bounded on or towards the north by the Grand Junction Canal the same two pieces of land to be used by the said company for the purposes of a station with yards while waiting boarding and unloading places warehouses and other buildings and conveniences for receiving depositing loading or keeping any Cattle or any Goods articles matters or things conveying or intended to be conveyed upon the said Railway or for any other purposes whatsoever connected with the said railway which the said Company shall judge requisite And the said Trustees further agree in person and of this power and authority given to them by the said Act to convey to the said Company and their successors the freehold and inheritance of the said two pieces of land (excepting and reserving unto the said Trustees all manorial rights whatsoever in over and upon the same) and it is agreed that the said Company shall by proper and effectual covenants provisions and conditions to be inserted in the said deed of conveyance or in some other Deed to the satisfaction of the said Trustees bind and oblige themselves in manner following (that is to say) - That the said Company shall at their own expense effectually fence off to the satisfaction of the said Trustees the above two pieces of land (except where bounded by the said canal) and also the two intended Roads between the said intended Station and the

25

Turnpike road and shall at all times thereafter keep the same fence in good order and efficient repair

That the said Company shall not erect or build any Inn Tavern or Public House upon the said pieces of land and that no houses cottages or [illegible] for habitation shall be erected thereon   [illegible] as may be necessary and requisite for habitation of the Officers Clerks Agents and [illegible] of the said Company employed at the Station

That the said Company shall and will use and appropriate the said two pieces of land or either of them for the aforesaid purpose of a station and not otherwise

That in the event of the said two pieces of land or either of them or any part thereof use respectively being discontinued to be used for the aforesaid purposes of a Station or being abandoned by the said Company for those purposes the said Trustees shall have the first offer of purchasing the land so discontinued to be used or abandoned as aforesaid and that the purchase money to be paid by the said Trustees for the same shall if the parties cannot agree be fixed by two indifferent persons one to be nominated by the said Trustees and the other by the said Company and that persons shall appoint an umpire if they cannot agree.

That wherever the expression "said Trustees" occurs in the foregoing Agreement it shall be construed to extend as well to the present trustees of the will of the said John Radcliffe as also to the future Trustees of the said will **In Witness** whereof to one part of this agreement with the said Trustees remaining the said Company have caused their Common Seal to be put and to the other part thereof with the said Company remaining the said Trustees have put their hand the day and year first above written.

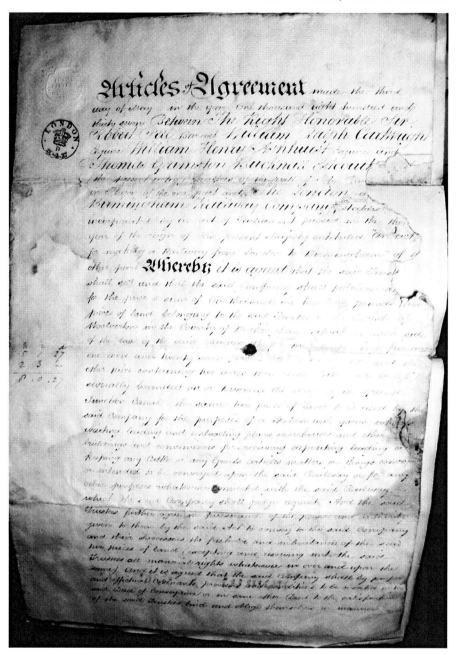

The first page of the Agreement

Road and River Diversion

The new viaduct necessitated a diversion of the course of the River Ouse. so that it had a straight passage underneath the viaduct and would therefore be less prone to silting up and flooding. This diversion took Mead Mill, one of the two Wolverton mills recorded in Domesday, out of commission. The Radcliffe Trustees may have concluded that one mill was sufficient to meet Wolverton's needs and that it should be closed down rather than be relocated and rebuilt. The mill was inhabited for a number of years after this before it was finally pulled down.

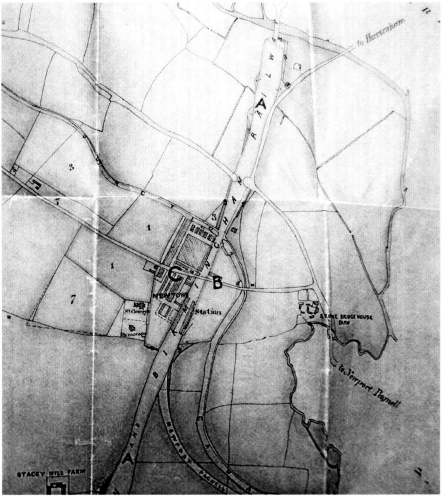

This section of an estate plan drawn up around 1847 for the Radcliffe Trustees shows the location of the former mill in the top right corner beside the embankment as well as the new river course.

28

The road to Haversham had to be diverted as well. The original intersection with the Old Wolverton road was to the west of the new railway bridge. As you can see the new road followed the line of the embankment before breaking to the north east and the river bridge. The line of this road was subject to several changes in subsequent years.

The Drawings of John Cooke Bourne

John Cooke Bourne (1814–1896) was an artist and engraver. He is best known for his lithographs showing the construction of the London and Birmingham Railway and the Great Western Railway. Each set of prints was published as separate books, and became classic representations of the construction of the early railways. Prints were often hand coloured for a vivid picture of events.

Wolverton Embankment

This view, on the next page, shows the massive earthworks under construction. 175 years later the embankment is part of the landscape but this drawing give some impression of the intrusive presence of the earthworks - all the more astonishing when you consider that this was done by hand before the invention of mechanical shovels. You can also see Mead Mill in this drawing. There was a mill recorded on this site in the Domesday Book but in 1838 it closed because the river had to be re-routed for the viaduct. The buildings survived as residential accommodation for another decade or so and then were demolished.

By February 1838 the two embankments still had to be joined to the ends of the viaduct, but the winter was then drawing to an end and Stephenson estimated that another eight weeks should see everything ready for the line to be laid. The section between Rugby and Denbigh Hall, a few miles south of Wolverton, was the last stretch to be completed, and finally, on 17 September 1838, the entire line was officially opened.

Wolverton Viaduct

The second Wolverton drawing shows the viaduct nearing completion. The arch scaffolding is in the process of being dismantled and already a train is testing out the line.

A Letter from Phillip Hardwick

Philip Hardwick (15 June 1792 – 28 December 1870) was an eminent English architect, particularly associated with railway stations and warehouses in London and elsewhere. Hardwick is probably best known for th Euston Arch and its twin station Birmingham Curzon Street, both designed a s symbolic entrances to the new world of rail travel.

The Doric Euston Arch at the old Euston station, was built at a cost of £35,000. It was, in the minds of many, demolished in the early 1960s. The gates of the arch are stored at the National Railway Museum in York. In 1994 the historian Dan Cruickshank discovered 4,000 tons, or about 60%, of the arch's stones buried in the bed of the River Lea in the East End of London, including the architrave stones with the gilded EUSTON lettering. This discovery has opened the possibility of a reconstruction of the arch.

The transcribed letter was written to the Board after he was sent to assess the Wolverton situation. At this late stage, only three months before the line was completed, there was no accommodation other than some wooden shacks for construction workers.

To Committee of Administration

London & Birmingham Railway

Russell Square

2nd July 1838

Gentlemen, Having agreeably to your desire inspected the Station and the several buildings proceeding upon it at Wolverton particularly with a view to ascertain the best position for the additional Cottages I beg to report that under all the circumstances, I consider the best site will be on the low Ground, between the large building and the Canal, where there is space for about fifty Cottages, each Cottage to contain two Stories, having one room on the Ground Floor and one above, - the latter divided if necessary – at the back of the Cottage a lean-to wash house and privy – with a small yard or garden – each Cottage to cost about £100 – the Timber Cottages at present standing on this Site to remain until they should become decayed or inconvenient from their situation. The Ground between the Railway and the first set of Cottages to be left for a wharf or a coal yard. The Cottages to be kept sufficiently back from the Canal to allow of Coals &c to be landed – A carriage road to be left on the west side of the principal Building with a sustaining wall in order that the new Cottages may be kept low, so as not to impede the light to the main Building. The Ground on the side of the Principal Building & which are now in progress I would recommend should be terminated in height at the second Storey and covered by a low projecting eave roof, instead of being carried up. Three Storey with Square parapets as intended, which would impede the light to the workshops. The Houses at the extremities of this row of buildings may be very beneficially converted into Shops & which I beg to recommend should be done.

As it is intended that the whole of the Ground on the opposite side of the Railway be appropriated for the purpose of Trade and Cattle – it will be necessary for the foundations now laid in for several dwelling Houses be taken up & the Bricks used in the new Works.

It appears that the Buildings at Wolverton have been executed under the immediate Superintendence of Mr Gandell the late Resident Engineer and the work executed by Mr Jackson a builder from London & during its progress is stated to have been measured by Mr Miller a Clerk of Mr Gandell's & priced according to a Schedule furnished by Mr Jackson and agreed to by Mr Gandell – this Schedule has been sent to me but until I have more particulars I can scarcely form an accurate opinion as to its fairness as it is formed upon a different principle from that upon which Architects are in the habit of contrascting.

The buildings which are now so much advanced must be of course finished at their Schedule Prices but it will be better for all the new Buildings to be done by contract at Specific Terms, avoiding as much as possible admeasurement, excepting when it may be rendered necessary by additions or omissions which occur in the progress of the work.

Mr Gandell having, I understand, been appointed to a more important situation in another Railway and having consequently quitted the Wolverton Station I considered it advisable in order that the Works should not be delayed to send down a Clerk of the Work, to have the charge of all the Building & the constant inspection of the Works – and to check all materials & Labour & who will I think be of material assistance in forwarding the work of the principal Building.

The advancement of all the Works & the arrangement of making up of the Accounts shall proceed under my own direction and the money now advanced & the balances paid under my Certificate.

All the new Works both in the Cottages & in the Station Buildings to be done by contract as specifications from Drawings & Specifications which are now preparing – but it will be important that all the materials that were used in the Cottages erected on the side of the embankment and taken down should be worked up in all these new Buildings – as well as the Bricks in the foundation of the Cottages which are now to be taken up.

I remain Gentlemen
Your most obed. Serv. **Phillip Hardwick**

33

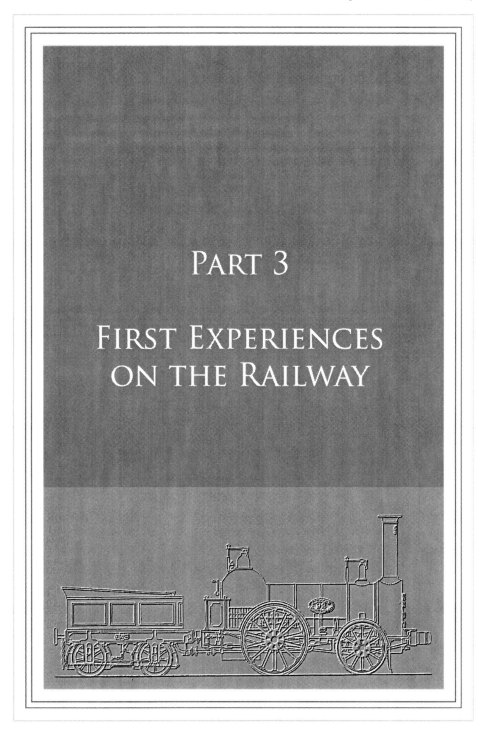

PART 3

FIRST EXPERIENCES
ON THE RAILWAY

IN THIS SECTION

Correspondence from early Travellers

The Times proved to be an effective vehicle for those wishing to express their dissatisfaction about their railway experiences. Although in its infancy, the postal service was swift and the railway of course did a great deal to improve the speed of communication.

- ### The Correspondence of George Phillimore

The Reverend George Phillimore of Willen was a prominent figure in the area. He was a Justice of the Peace as well as a clergyman. His letters detail some early frustrations with rail travel before the line was complete.

- ### A Letter to the Morning Post

This letter was composed in the middle of the night by a frustrated traveller who, travelling on the first train from Euston to Birmingham, had missed his connection for his train to Liverpool on the Grand Junction Railway.

PART 3: FIRST EXPERIENCES ON THE RAILWAY

The Correspondence of the Rev. George Phillimore of Willen

Wolverton was late opening. The middle section of the line was a very difficult construction project. problems were encountered with the Ouse viaduct, digging into hard rock at Blisworth and considerable delay in completeing the Kilsby Tunnel The directors therefore resorted to the expedient of running trains from London to Denbigh Hall bridge, where the line intersected the Watling Street and transporting passengers by coach to Rugby. Some passengers were not happy, as may be illustrated by this correspondence in the Time from the Reverend George Phillimore of Willen.

Wednesday September 12th 1838

TO THE EDITOR OF THE TIMES

Sir, - Knowing the interest you feel upon with all subjects connected with the welfare and convenience of the general public, and relying upon the laudable zeal which your able journal has aways exhibited in the exposure of abuses, I hope you will allow me, through its valuable medium, to state an occurrence which happened to me a short time ago when travelling by the London and Birmingham Railway. That company is, by act of Parliament, invested with such extraordinary powers, that it appears to me to be the bounden duty of every individual who may suffer from any abuse of these powers to make known his grievance, in order that such details may operate as a stimulus upon the Legislature to interfere in behalf of the public, before the monopoly becomes complete by the extinction of stage coach travelling. Permit me, then, to lay before you a correspondence which has taken place between myself and the directors, upon the occurrence in question, which seems to me will speak for itself.

Willen Vicarage, August 25, 1838.

"Sir - I beg to call your attention to the following occurrence which happened to me upon my arrival at Denbigh hall, by the train which left London at 3 p.m. this day, I was accompanied by a lady, who was going to

Northampton. In order to secure an inside place by the coach from Denbigh hall, she had, at my suggestion, booked and paid for, at the Golden Cross office, Regent Street, a place in the first class carriage from London, (getting in, however, at Watford,) and an inside place by the Northampton coach. At the Golden Cross a receipt was given for the payment of the fare thither. The receipt was objected to at first by the Watford bookkeeper as not being a proper ticket, but upon conferring with the guard of the train, he admitted it to be so, and directed him to pass my companion to Northampton. Upon arriving at Denbigh hall, however, great was my surprise to find this ticket refused by Simcox, the check taker, who affected entire ignorance of its purport. When upon this I assured him that everything was right, and that the guard knew it to be so, he answered me insolently, that he must be paid. Wishing to avoid any unpleasant altercation occurring before my companion, I desired her to pass on, and told her that I would settle the matter. Again, I assured the check taker that the fare was paid, when, upon endeavouring to pass him, I was immediately seized and collared by him! In spite of my remonstrances I found it necessary to break away, and whereupon he called policeman 73, to whom he gave me in charge. I offered an explanation to the policeman, which he peremptorily refuse to hear, and proceeded to collar and detain me. I begged the guard might be summoned, who at once corroborated the truth of my statement. I then offered to go away, but was again seized and detained by the policeman. I asked Simcox his name, who refused to give me his name, as did the policeman. I lose no time in forwarding my complaint, and beg your immediate attention to it, and a satisfactory redress for the insolence and assault on the part of two of your servants, forebearing to make any comment, however obvious, upon the above statement. I ought to add, that before they would release me, 7s. was demanded, which I gave, in payment of the fare; and also that the coachman of the Northampton coach showed me my companion's name on the way-bill, as booked from London to Northampton, and at my desire showed it to Simcox, who persisted in refusing to acknowledge its correctness.

"I am, Sir, your obedient servant,
GEORGE PHILLIMORE"
To the Secretary of the London and Birmingham Railway."

To this letter I received the following reply:-

"London and Birmingham Railway-office,
Euston-square, August 30, 1838.

"Sir - Having laid your letter of the 25th inst. before the committee of management, and a rigid investigation into all the circumstances connected with your complaint having been immediately instituted, I am instructed to express to you the regret of the company that you should have been exposed to so much annoyance, and to apologize on their behalf for any failure in proper respect to a passenger on the part of any of their servants. I am desired to add, that the immediate cause of this annoyance is the conduct of persons connected with the intermediate coaching, and that this evil will cease on the general opening on the 17th September.

"I am, Sir, your obedient servant,
"R. CREED, Sec.

"To the Rev. S. Phillimore, Willen, Newport Pagnel."

Richard Creed was the London Secretary of the Company, which also had another, in the person of Captain Moorsom, at its Birmingham office. Both held important and influential positions and may have had a larger role than Company Secretaries have today.
The Rev. Phillimore was not to be fobbed off so easily.

To this I replied as follows:-

"Sir, - I received duly your letter this morning; and, in reply, must candidly confess that the explanation which it contains is anything but satisfactory. I do not find therein any mention made whatever of reprimanding, or in any way punishing the servant or servants, through whose negligence I was insulted, but the onus of the transaction is attempted to be laid upon the 'persons connected with the intermediate coaching,' a statement which, if borne out by the fact, would only throw additional blame upon the Company for their bad arrangement. Again, a point upon which not only myself, but the public at large, are materially interested, viz., the extortion of money for a fare already paid, and the restitution of money obtained under circumstances of personal violence and detention, is passed over, after the "rigid investigation," which you state to have taken place, in total silence. What am I to infer from this! Either that the 'investigation' was a very loose one indeed, or that the money will not be restored. Once more, therefore, I am compelled to draw your attention to the following points: - 1. The book-keeper of the Golden Cross, which in your advertisement in The Times is stated, under the signature of your two secretaries, to be one of the offices where places may be taken, gives my servant a ticket, upon payment of the fare, to Northampton. The servant expressly asks whether that ticket is sufficient. The book-keeper replies that it is.

"The book-keeper at the Watford station, after learning from the guard of the train that the lady whom I accompanied was duly booked, desires the guard to pass her on to Northampton free. Now, in the case of this ticket being an improper one, the book-keeper at the Golden Cross (which is expressly declared by public advertisement as an authorized office for booking places on the railroad), was to blame for not giving a correct one; and in any case, it was the duty of the Watford book-keeper to have seen that she was provided with a proper ticket before allowing her to enter the carriage. But it does not appear that the latter ever doubted the validity of the ticket, for he took the assurance of the guard, and declared the fare paid to Northampton. Is the negligence of your servants to be made the cause of assaulting and wounding the feelings of a gentleman, and the justification of committing extortion upon a passenger? Are the public to suffer from the carelessness or faulty arrangements of the railroad directors? Are they to believe the advertisements which they seen in the newspapers, signed by the two secretaries, stating that there are certain offices in London where places may be booked? Are they made to be liable, when travelling by your conveyances (whether male of female, gentle or simple) to be assaulted, and, for a mistake clearly committed through the fault of your Company's servants, to be collared, and given in charge to a policeman, and, in order to obtain release, to be compelled to pay a fare over again; and in the end, when demanding redress for such usage, to be soothed with the satisfactory intelligence that "the immediate cause of this annoyance lies with the conduct of persons connected with the coaching department?" I am very much mistaken if the public at large will put up by these things, or if the Birmingham Railroad Company will find their interests advanced by such occurrences, and such unsatisfactory replies to just complaints. I am not of a litigious disposition, but every man possesses feelings which may properly be called into exercise on certain occasions, and these urge me, as well upon public as upon personal grounds, to seek a satisfactory redress and a restitution of my money, both of which demands I feel confident the committee will see the propriety of acceding. I await your immediate reply, &c.

"I am, Sir, your obedient servant,
"GEORGE PHILLIMORE.
"To the Secretary of the London and Birmingham Railroad."

In this account we only get the Reverend Phillimore's side of the story. It does appear that he had foundation for his grievance. In mitigation for the railway, these were very early days. All the jobs were new and the Company was unable to call upon any experience. Men were hired, provided they could read and write, on the recommendation of a worthy and respectable person. Staff training, as such, was probably non-existent, but employees were provided with written instructions, or "rules" for the execution of their duties. They probably did not cover situations like this.
To this the Committee replied thus:-

"London and Birmingham Coaching department,
"Euston-station, Sept. 3, 1838.

"Sir,- I am instructed to transmit to you a Post-office order for 7s., in reference to your communication of the 31st ult., to which the Secretary has replied.
"I am, your obedient servant,
"A. BAGSTER."
"Rev. G. Phillimore, Willen.

It appears that Richard Creed was unwilling to waste any more time on this country parson and delegated the matter to someone in his office. He may have overlooked the fact that Phillimore was 7s. out of pocket from the first letter. His instruction must have been, "pay the man and close the matter."

Reverend Phillimore still had bruised feelings.

Upon submitting the above to a respectable solicitor, it appeared that my only remedy for the above treatment was to enter an action against the Company - a remedy expensive and unsatisfactory; but upon mature consideration, reflecting that my end being not merely of a personal nature, but to expose the abuses of a great monopoly to the public, with the hope that the Legislature will interfere to put a stop to them, I deemed that the most effectual means of doing so would be to obtain a place for the above correspondence in the columns of your widely circulated journal.

I am, Sir, your obedient servant,

GEORGE PHILLIMORE.

Willen Vicarage, Sept. 7.

A few days later, Phillmore threw in the towel and resorted to the publicity he could get by publishing the correspondence in *The Times*. I don't know if this is one of the first instances of a customer using the press as a means of getting attention to a perceived wrong but it is probably an early instance. The London and Birmingham Railway is an early instance of a large public company providing service to the public and entering into a new type of customer service. If Josiah Wedgewood sold you a flawed piece of china it could be replaced, and no harm done. If a railway traveller had a poor travelling experience it could not easily be put right. The L & BR plainly did not know how to deal with such situations as experienced by George Phillimore and by today's standards would fall very short. Everybody was learning.

*

Two weeks later the line was fully open. Inevitably there were teething troubles as the following sardonic letter, written the day after the line opened show, This traveller, taking at face value the assurance that he could travel from London to Birmingham and easily catch the GJR train leaving Birmingham at ten past three, In the event he missed it by a few minutes and had to spend the night in a waiting room, where he penned this letter.

Morning Post Thursday 20 September 1838

TO THE EDITOR OF THE MORNING POST

Sir - I am induced to trouble you with this, in order that we may serve as a warning to other unfortunate individuals who are addicted to reading advertisements, and, what is worse, believing them. Wishing to save time and leave Liverpool very early this morning, we got into the mail carriage of the London and Birmingham Railway at half-past eight on Monday evening, with the assurance that we should go on direct by the railway throughout, and be forwarded by a corresponding train of the Grand Junction Railway, leaving Birmingham at eight minutes past three - so ran the printed papers we were shown. There appeared to exist some hitch on the road to give us a hint of our fate, as when we left the station at Euston-grove there were two trains due which should have arrived at six and eight o'clock. We had the pleasure of seeing them arrive while our train was making some abortive attempts at starting. At length we were off; but our engine appeared to be affected with asthma, for it wheezed, and puffed, and dragged the weary length of the train behind it at a miserable pace, every now and then bringing up to get breath. At length we arrived at Wolverton, having previously been left by the lucky real mail passengers to be dragged along with more certainty and nearly as fast by "real horses" as they say at Astley's. At Wolverton another engine was attached, which seemed to be willing to make up for some of the time lost by its predecessor, which had managed to convey us from London to Wolverton in about four hours. leaving Wolverton at half past twelve, we pulled up at Rugby for a few minutes to land passengers and get a cup of coffee. There the inspector comforted us with the sight of a placard, on which was printed assurance that a train would leave Birmingham for Liverpool at eight minutes past three, and that though we were one hour behind our time yet we should be in plenty of time for the said train. We arrived at Birmingham at ten minutes past three, where, in the yard of the railway station, we found two cars and two omnibuses, the drivers of which hastened to assure us there was no conveyance to Liverpool until six in the morning, ergo, that we had better for their good go to Birmingham, and allow them the felicity of jolting us down, and then, having inhaled the fragrance of the Birmingham inns, allow ourselves to be jolted up again. Of course we expected that there would have been some person in authority who could have informed us how to proceed in securing immediate conveyance to Liverpool, but although there were policemen and other badged gentry in abundance, they all told different tales. one assured us there was a train at eight minutes to three, another at four, and another at six; but, alas! in the midst of this enchanting vision of speedy exit from the land of "buttons and brass." a grim-visaged gentleman, in uniform and badge, bearing a lantern, approached, and told us no one knew anything but him (I set him down immediately for a full-grown Brummagem denizen); that there was no train a all for Liverpool until six in the morning; that the Grand Junction Railway station would not be open; and that if we went there we should stand in the street. However, after telling them our minds pretty freely, we persuaded the omnibus driver to deposit us, luggage and all, at the Grand Junction railway station. they have, like good Samaritans, opened their doors to us, and even gone the length of hospitably making some fire. However, here we are until six o'clock, when, in spite of my hurried over my affairs in town, to get away in time to leave Liverpool by an early vessel, I shall arrive there just too late because I believed the advertisements of the London ad Birmingham Railway Company, who have delivered us here, 114½ miles, in six hours and forty minutes. However good the arrangements may be on the lines taken individually, the public have a right to demand that there should be such an understanding between the railway companies as shall insure the rapidity of travelling and certainty of arrival, which they have the power to guarantee, and also that the private understandings of individuals should not interfere with the public good.

Birmingham, Tuesday morning, Four o'Clock.

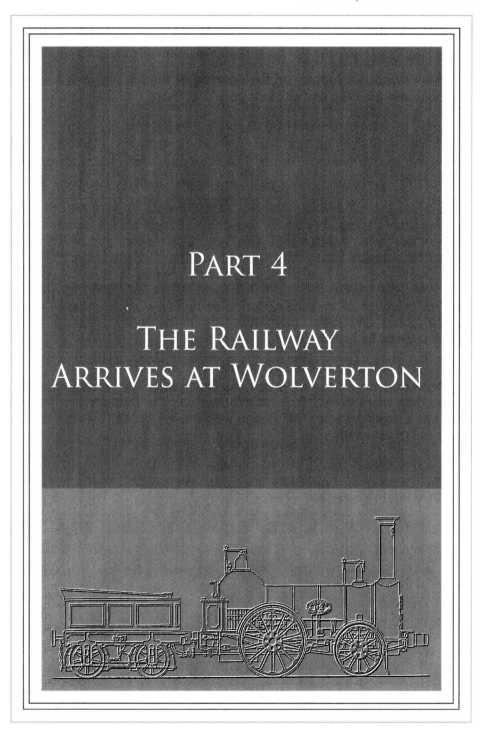

PART 4

THE RAILWAY
ARRIVES AT WOLVERTON

In this section:

- Opening Day

Two newspaper reports on the opening of the full line: one from the *Northampton Mercury* and another from *The Times*.

- A timetable from 1839

This was posted at The Swan at Newport Pagnell. The station is described as "Wolverton Central", i.e. central to Newport Pagnell and Stony Stratford.

- The First Railway Station

This was a temporary set of wooden buildings on the embankment north of the canal. It lasted two years until a permanent station was built south of the canal.

- Minutes from Some Board committee Minutes

This selection of minutes describe some of the early decision made by the board and their committees about the many pressing social needs created by the new town.

- The Second Railway Station

Wolverton was an important resting place for early travellers. Passengers were given 10 or 15 minutes and the Refreshment Rooms here did a roaring trade. The station was moved in 1881 when the loop line was constructed.

- The Binns and Clifford Survey of 1840

The Birmingham surveyors prepared this plan in 1840. Wolverton was very much a work in progress.

- Francis Wishaw. The Railways of Great Britain and Ireland

Francis Wishaw was a prominent engineer who left this detailed report of Wolverton in 1842.

- Extract from the Railway Journal

John Herapath was an early and enthusiastic railway writer. This is his Wolverton report.

- Extract from the Census of 1841

- The Directors and Chief Officials

PART 4: THE RAILWAY ARRIVES AT WOLVERTON

Opening Day

Most newspapers of the day reported the event, although in many cases they all printed the same copy. Here are two reports, one from the *Northampton Mercury*, with a natural bent towards Northampton's interest and a more complete report from *The Times*.

Northampton Mercury, September 22nd 1838

The London and Birmingham Railway was opened the entire distance on Monday last. A special train left the Euston station at twenty minutes after seven, conveying G.C. Glyn Esq. Chairman of the Board of Directors, Mr. Stephenson, the engineer, Mr. Berry, the contractor for locomotive power, Mr. Creed, one of the secretaries, and Messrs. Calvert, Baxter, and other Directors and officers of the Company. The Duke of Sussex and suite also travelled by this train as far as Rugby. The first public train left London shortly after eight o'clock, and consisted of nine first-class, two mail, and four gentlemen's carriages, conveying altogether nearly two hundred passengers. There was a vast assemblage of persons, on the numerous bridges which cross the railway on the London side of Primrose Hill, to witness the departure of these trains, who testified their interest in the success of the undertaking by cordial cheers. The first train to London left Birmingham at seven in the morning. No accident whatever occurred during the day, which fortunately was a remarkably fine one. The stations from London to Birmingham are as follows:-

At Harrow, distance 11½ miles; Watford, 17¾; Boxmoor, 24½; Berkhampstead, 28; Tring 31¾; Leighton, 41; Wolverton, 52½; Roade, 60; Blisworth, 63½; Weedon, 69¼; Crick, 75½; Rugby, 83¼; Brandon, 89¼; Coventry, 94; Hampton, 100½; Birmingham, 112¼.

The opening of the entire line, by no means, however, implies the completion of this stupendous work. there is still an immense deal of labour to be performed along the whole distance, and especially at the newly opened portion. many of the stations also are yet very incomplete;

that at Blisworth consisting of little more than a wooden shed, and a tremendous flight of wooden steps.

In this town and neighbourhood the event excited the greatest interest, and the new portion of the line was thronged by thousands. Every thing in the nature of a conveyance which the town could afford was in requisition at an early hour. Denbigh Hall, which so recently enjoyed the distinction of a "Terminus," is now completely deserted, not even being a "Station." The wooden offices are closed; the wattled stabling, and the greensward stable-yard, lately so crowded, are empty; and the tarpaulin-roofed "Hotel" no longer offers to the night traveller the attraction of a cheerful bonfire, and a boiling kettle without, and a ham sandwich and a hot cup of coffee within. All is tenantless, and the train shoots by unheeded. The glory of Denbigh Hall is departed.

We are glad to observe that in the useful little table of distances and fares, published by the Company, Blisworth is marked as a First Class Station.

The Times, the edition of September 18th, 1838.

Yesterday was the first day that the complete line of railroad from the London to the Birmingham terminus was opened. The portion of the road which was traversed for the first time on this occasion was that which extends between the old station at Denbigh hall and the station at Rugby. The station at the former place now no longer exists; but there are on this extent of 35 miles stations at Wolverton, Roade, Blisworth, Weedon, and Crick. The first train started from the Euston square station at 7 o'clock, having in the carriages the proprietors of the undertaking and their friends. It was said in Birmingham that they accomplished the whole journey in four hours and a half. The next train, which was open to the public, left Euston square station at 10 minutes after 8 o'clock, but did not get fairly under weigh with the steam engine until 25 minutes past 8. The train reached Birmingham by the Birmingham clocks at the terminus at two minutes to 2. Watford was reached in 33 minutes from the Euston station. The train halted there three minutes. Tring was reached in 73 minutes, and the train halted four minutes and a half. Wolverton, the first new station, was reached by 28 minutes

48

past 10, the the train halted 25 minutes. At this place a great crowd of persons were assembled, and preparations were made for a rural feast and celebration of the opening of the line. Roade was reached at 17 minutes past 11, the train stopped 10 minutes at this station, which is 60 miles from London. Weedon, which is nine miles further, was reached at 7 minutes to 12 o'clock, and Rugby, which is 83 miles miles from London, at half pat 12. The train stopped here 8 minutes. Coventry was reached at six minutes part 1 o'clock, and here the train remained for 15 minutes. The next place was Birmingham. The portion of the line just opened, from Denbigh hall to Rugby, appears to be equally good with any other part of the road. It is this division of the road, shortly before entering Rugby station, that the trains pass through Kilsby tunnel. It has been asserted that this tunnel fell in during the boring of it, but it is not the case. It is one of the most extraordinary pieces of road in the whole line. The length of this tunnel is 2,400 yards in length, and does great credit to the skill of Mr. Foster, the engineer by whom it has been completed. The train which left Birmingham for London at half past 12 was delayed, by some means or other, on the road for nearly two hours, in consequence of which, the train next in succession, which left Birmingham at half past 2, was delayed almost two hours when almost close to Euston station; this last train arrived in London about 20 minutes to 10, instead of a quarter past 8, the hour stated for arriving in public announcements. It does not appear that any accident whatever occurred on the road; indeed so excellent were the arrangements, that the possibility of accident was provided for in every way that could be imagined.

There is a lot of detail about time in this report and a journey of four and a half hours was a long one. But a journey of this length, which would hitherto have taken 10 hours by the fastest stagecoach was an amazing phenomenon to those early Victorians. The journey from Euston to Wolverton took three hours and the passengers would have needed the 25 minutes to relieve themselves at the new station. Obviously someone had taken the trouble to organize a "rural feast". The moment signalled a great change for Wolverton.

A timetable from 1839

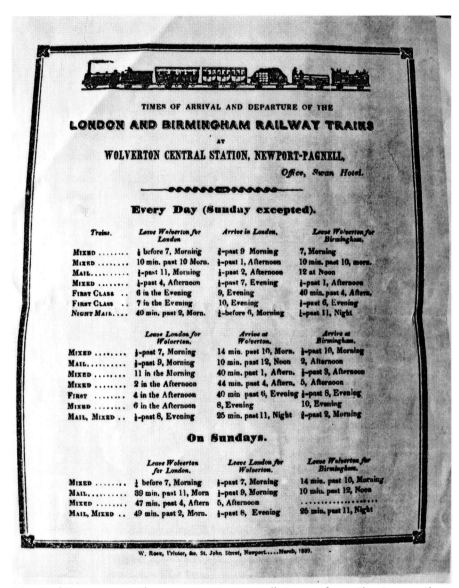

TIMES OF ARRIVAL AND DEPARTURE OF THE

LONDON AND BIRMINGHAM RAILWAY TRAINS

AT

WOLVERTON CENTRAL STATION, NEWPORT-PAGNELL,

Office, Swan Hotel.

Every Day (Sunday excepted).

Trains.	Leave Wolverton for London	Arrive in London.	Leave Wolverton for Birmingham.
MIXED	¼ before 7, Morning	½-past 9 Morning	7, Morning
MIXED	10 min. past 10 Morn.	½-past 1, Afternoon	10 min. past 10, morn.
MAIL	½-past 11, Morning	½-past 2, Afternoon	12 at Noon
MIXED	½-past 4, Afternoon	½-past 7, Evening	½-past 1, Afternoon
FIRST CLASS	6 in the Evening	9, Evening	40 min. past 4, Aftern.
FIRST CLASS	7 in the Evening	10, Evening	½-past 6, Evening
NIGHT MAIL	40 min. past 2, Morn.	½-before 6, Morning	½-past 11, Night

	Leave London for Wolverton.	Arrive at Wolverton.	Arrive at Birmingham.
MIXED	½-past 7, Morning	14 min. past 10, Morn.	½-past 10, Morning
MAIL	½-past 9, Morning	10 min. past 12, Noon	2, Afternoon
MIXED	11 in the Morning	40 min. past 1, Aftern.	½-past 3, Afternoon
MIXED	2 in the Afternoon	44 min. past 4, Aftern.	5, Afternoon
FIRST	4 in the Afternoon	40 min past 6, Evening	½-past 8, Evening
MIXED	6 in the Afternoon	8, Evening	10, Evening
MAIL, MIXED	½-past 8, Evening	25 min. past 11, Night	½-past 2, Morning

On Sundays.

	Leave Wolverton for London.	Leave London for Wolverton.	Leave Wolverton for Birmingham.
MIXED	¼ before 7, Morning	½-past 7, Morning	14 min. past 10, Morning
MAIL	39 min. past 11, Morn	½-past 9, Morning	10 min. past 12, Noon
MIXED	47 min. past 4, Aftern	5, Afternoon	....
MAIL, MIXED	49 min. past 2, Morn.	½-past 8, Evening	25 min. past 11, Night

W. Rose, Printer, &c. St. John Street, Newport.....March, 1899.

Most of the new stations were at some distance from the towns they were expected to serve. Four miles obviously presented no impediment to Newport Pagnell travellers who appear from this poster to regard Wolverton as *their* station.

The First Railway Station

As you can infer from the comment about the Blisworth Station in the *Northampton Mercury*, the first stations were hurriedly-built wooden constructions. Wolverton was no exception and a station was built on the north side of the canal on the embankment. It was replaced within two years and the only record of it is a rather small line engraving.

WOLVERTON STATION.

The view is from the east side. The building on the left may have been the Pumping House for the works' water supply. To the north of the bridge the artist has drawn some rather rudimentary buildings. The V-shaped wall (if that is what it is) appears to cover steps up to the station.

The plan drawing on the next page will give a better indication of the placement of the station.

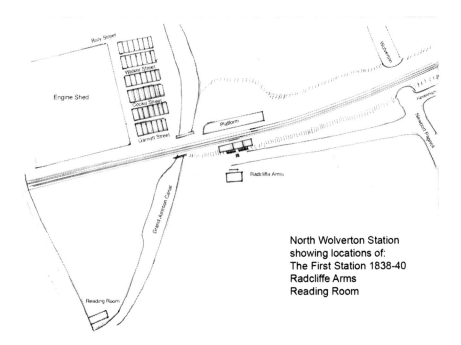

North Wolverton Station
showing locations of:
The First Station 1838-40
Radcliffe Arms
Reading Room

Minutes from Some Board Meetings

London and Birmingham Railway Board Meeting

Euston 20th December 1839

The Chairman recommended that Schools be established at Wolverton for the use of the Workmen in the Company's employ.

1121: Resolved that the proposition be approved and adopted and that the Chairman, Deputy Chairman and Mr. Sturge be a Special Committee for carrying the arrangement into effect.

Locomotive Power Committee

Wolverton 9th January 1840

Read Mr. Bury's Report of the 30th Dec.

It appearing that the number of persons employed in the Locomotive Department at Wolverton is 203 including 4 at salaries of £150 a year and upwards, that the number of houses and Cottages for their accommodations is 62 the chief part of which are let at 3/6 and 4/- per week and a few at 5/- and 7/- per week and that it is extremely difficult to keep good workmen there unless there be suitable residences for them and their families.

The Committee recommend that the Trustees of the Radcliffe Estate be treated with without delay for the purchase of sufficient ground to build 50 cottages at a cost not exceeding £80 to £100 each, and five houses for superior officers at a cost not exceeding £1000 for the five houses and that these Cottages and houses be built for the Company by contract. It being expected that 10 per cent will be readily obtained on the outlay.

That Mr. Bury be requested to consider whether a row of Cottages may not be built on the land belonging to the Company along the Bridge embankment on the east side of the Railway as part of the accommodation wanted.

That a piece of ground be also applied for to the Radcliffe Trustees in a convenient situation and at a reasonable rent or lease for the purpose of its being divided into small gardens and underlet, at such rates as will cover expenses, to the men employed at Wolverton.

London and Birmingham Railway Board Meeting

With reference to the report of the Locomotive Committee,

1148: Resolved that the Chairman be authorized to continue his negotiation with the Radcliffe Trustees for additional Land at Wolverton.

Coaching and Police Committee

Euston 15th January 1840

Resolved that it be recommended to the Board to build the new Station and Refreshment Room at Wolverton on the south side of the Bridge if some land can be procured from the Radcliffe trustees and that with some required alterations the plans formerly arranged and contracted for with Mr. Jackson be recommended for adoption.

The Second Railway Station

The first station was a temporary building and it was always anticipated that a more permanent building would replace it. Most believed, and certainly the unlucky builders of The Radcliffe Arms in 1839, that the original site would continue to be used. They had good reason to believe this. The turnpike road which we now call the Old Wolverton Road was close by and newer roads to the south simply did not exist. It was therefore a surprising, even shocking news that the railway board intended to relocate the station in 1840. There was more land certainly and passengers no longer had to climb steps up a steep embankment, but the company had to build a new road and a canal bridge to provide access.

Wolverton was deemed to be of sufficient importance to be graded a "First Class" station, a distinction it shared with Tring, Blisworth and Rugby. It accommodated the famous Refreshment Rooms (described later) and a hotel was planned for the station. It was never built, as speedier and more reliable engines brought about Wolverton's decline as a mandatory stopping point.

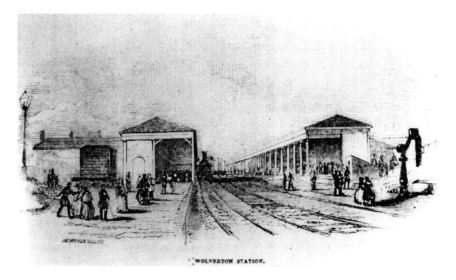

WOLVERTON STATION.

The Binns and Clifford Survey

In 1840 the Birmingham Surveyors, Binns and Clifford, were instructed to draw up a plan for Wolverton. the northern part was already complete, but the southern section was in progress and was not actually completed until about 1847. Their plan, showing drainage and water supply is the earliest surviving plan of Wolverton station.

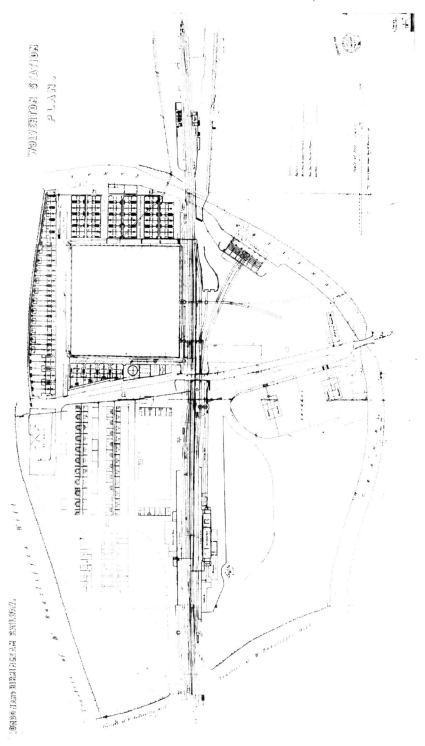

WOLVERTON STATION PLAN.

LONDON AND BIRMINGHAM RAILWAY.

The Description of Francis Whishaw

Francis Whishaw (1804 - 1856) was a Civil Engineer and wrote *The Railways of Great Britain and Ireland*, published in 1842, a monumental work containing an exhaustive general and technical account of nearly all the railways open or under construction at that time. As Secretary to the Society of the Arts, Francis was also one of the prime movers behind shaping the Great Exhibition of 1851. This account, which has been extracted from his book, provides us with an engineer's view of the new town of Wolverton.

LOCOMOTIVE ENGINE-DEPOT, GOODS-DEPOT, AND PASSENGER-STATION AT WOLVERTON.-

The buildings lately erected at Wolverton, as the principal station for the locomotive engines, form, perhaps, one of the most complete establishments of the kind in the world. The site of this establishment is on the left side of the railway, at a distance of about 52 miles from the London terminus, and 59 ½ miles from that of Birmingham, having a frontage on the Grand Junction Canal. The buildings, which are of plain but neat design, and constructed chiefly of brick, surround a quadrangular space 127 feet wide by 216 feet deep, the entrance to which is under an archway in the centre of the principal front. The whole length of the buildings is 221 feet, the depth 314 feet 6 inches, and the height 23 feet; the main walls are 2 bricks in thickness. Besides the central gateway, which is 12 feet 6 inches in height above the rails, there are two side-entrances: the one to the large erecting shop, the other to the repairing-shop.

The erecting-shop is on the right of the central gateway, and occupies one half of the front part of this building. It has a line of way down the middle, communicating with a turn-table in the principal entrance, and also with the small erecting-shop, which is on the left of this entrance. Powerful cranes are fixed in the erecting-shops for raising and lowering the engines when required.

Contiguous to the small erecting-shop, and occupying the principal portion of the left wing, is the repairing-shop, which is entered by the left gateway. One line runs down the middle of this shop, with nine turn-tables, and as many lines of way at right angles to the central line. This shop is 131 feet 6 inches long and 90 feet wide, both in the clear, and will hold eighteen engines and tenders, or thirty-six engines. It is lighted by twenty four windows, reaching nearly to the roof.

In the same wing, and next to the repairing-shop, is the tender-wrights' shop, having the central line of way of the repairing-shop running down its whole length, with a turn-table and cross line, which runs quite across the quadrangle, and intersects a line from the principal entry to the boiler-shop in the rear of the quadrangle.

The remainder of the left wing is occupied by a room for stores on the ground-floor, with a brass-foundry and store-room over; and the iron-foundry, which extends to the back line of the buildings.

The right wing contains the upper and lower turneries, each 99 feet long and 40 feet wide; the upper floor being supported in mid-line by nine iron columns. There are fourteen lathes in the lower, and eight in the upper turnery. The fixed pumping-engine house is also in the right wing, occupying the central portion thereof, and measuring 26 feet 3 inches by 19 feet 6 inches. There are two engines, each having a 14-inch cylinder and 4-feet stroke, and worked with from 35 lbs. to 40 lbs. pressure; the fly-wheels making twenty-four revolutions per minute. The boilers are placed in a sunk area in front of the engines, and separated there from by a 9-inch wall. The water is pumped from a well in the centre of the engine-house; this well is of elliptical form, the transverse and conjugate diameters of which are respectively 11 feet 6 inches and 8 feet 2 inches. The brickwork is 9 inches thick, and the whole depth of well 93 feet. At the bottom of the well are two tunnels, running north and south, each extending 33 feet from the well. These tunnels are 8 feet wide, 8 feet 6 inches in extreme height, and 6 feet to springing of segmental arch; the brickwork is 13 ½ inches in thickness. The two pumps are each of 7 inches diameter. There are two tanks to receive the water from the well: the one above the engine-house having a capacity equal to 2590 cubic feet, or 15,540 gallons; and the other 3850 cubic feet, or 23, 100 gallons. This latter tank is over the gateway.

Besides pumping water for the establishment, and giving motion to the lathes and other machinery, these engines have another duty to perform, which is that of working the blowing-machine. The blowing cylinders are fixed on a floor above, and immediately over the engine-cylinders, are 3 feet in diameter, and are worked by the same piston-rods, having a 4-feet stroke. The air is admitted at the top of the blowing-cylinder by a pipe communicating with a vertical cylinder, 10 inches in diameter, which is carried out above the roof. The 9-inch blast-pipe passes from the top of the cylinder, on the opposite side to that in which the air is admitted, and runs down to the level of the smithy, to blow the numerous fires which range along the sides and ends.

The smithy occupies the north-west angle of the building, running partly
down the right wing, to the extent of 137 feet 3 inches, and joining the engine-house, and partly along the back portion of the building, to the extent of 76 feet. It contains eighteen single, and three double hearths.

The remaining space of the back portion of the buildings is occupied by a joiners'-shop, with store-room and pattern-shop above, the hooping-furnaces, and a boiler-shop. In the boiler-shop there are two hearths; and, communicating with the machinery worked by the engines above described, are two drills and a punch.

The lodge, superintendent's office, and drawing-office, are in a building within the quadrangle, and close to the principal entrance.

The various departments of this establishment are warmed by steam, issuing through cast-iron pipes laid in channels, paved over, and furnished with proper ventilators.

On each side of this extensive structure there is a street running down to a wider street at back, which is 40 feet in width, including the footpaths. In the left street are the gas-works, and eight cottages of two stories for the workmen. Between the street on the right and the canal, other streets run down at right angles. In the principal street, which is at the back of the locomotives' building, there are six houses of three stories, for clerks and foremen; twenty-two of two stories ; and eight with shops on the ground floor. From the main street there is a communication with the high road, which passes over the railway to the south of this station.

In front of the locomotives' building there are four lines of way; the main double way being in the middle, with an intermediate space of 6 feet 5 inches, the whole width of way being about 60 feet. On the side-line next the building are two engine-races, or pits, 3 feet 9t inches wide, and 2 feet 4 inches deep from level of rails. A grating is fixed in the bottom of each race, to let off the water from the engines, when required, into a proper drain below. On the cross lines, which are in communication with the locomotives' establishment, are six turn-tables; two of which are in front of the carriage landing, which is on the east side of the railway. In this carriage-wharf or landing there are two docks or recesses, each 9 feet 2 inches wide, 5 feet deep, and 3 feet 84 inches high, with proper indents in the coping to receive the buffers and chains. To support this wharf four needle piles of oak are driven at a distance of about 10 feet 6 inches from the back of the wall, and between these piles and the elm planks, which are close to the wall, strutts are introduced 10 feet in length. The whole width of this landing is 28 feet 6 inches, and it is run out with a proper slope leading from this station to the main road.

Fronting the canal, and on the east side of the railway, is the goods warehouse, which is furnished with a double way, forming a communication with the main line. There are two lines of way running down the length of the warehouse between two stages or platforms, 15 feet wide and 4 feet high. On these stages are cranes for raising or lowering goods from or to the canal-barges or railway-wagons. Beneath the front stage is a coal-store, with six loop-holes next the canal. This building is lighted by four skylights in the roof, which is slated, and projects over part of the canal to protect the barges in bad weather.

The temporary passenger-station is on the north side of the canal.

Here the trains are allowed to stop ten minutes, provided they arrive in proper time, for the purpose of allowing passengers time to take refreshment.

Every engine with a train from London or from Birmingham is changed at the Wolverton station, which answers the double purpose of having it examined, and of easing the driver and stoker. We consider even fifty miles too great a distance to run an engine without examination; and have seen on other lines the ill consequences arising from the want of this necessary precaution. We should prefer about thirty miles' stages, when it can be managed.

Extract from the Journal of John Herapath Editor of the Railway Magazine

THE WOLVERTON STATION

One of the great lions on the road is this establishment, 52½ miles from London and 59¾ miles from Birmingham. I could not permit myself to go upon the line without inspecting the magnificent works at this place, particularly as an offer to see them was tendered to me by Mr. Creed. A gentleman by the name of Stafford[1], by the direction of Mr. Bury, went with me over the works, and the little town which has been reared around them.

Within about three years a large factory, standing on 1½ acre of ground, employing abut 300 mechanics and labourers, together with 200 houses, occupied by about 900 souls, a place of worship with a resident clergyman, and schools for boys and girls, has sprung into existence; and a square for a market place is already laid out.

The houses are built by the Company, and let at remarkably reasonable rents, to their employees. Houses, which in the neighbourhood of London would readily fetch from £16 to £25 a year, are let from 1s.6d. to 2s.6d. a week. but there are some let at 6s a week, if not higher. The object of the Company is not to make a profit, but simply to pay the interest of the outlay. Of course this is equivalent to giving the men so much higher salaries. Streets, though at present not evidently under the care of Macadam, are made, and shops already exist, and the place has even now the air of a busy little town. The gardens are not attached to the houses, but are altogether, in what may be called the suburbs; and as they are separately rented, everyone who has taste that way, enjoys himself with his own garden.

Close to the west end of the town, which circumscribes three sides of the factory, runs the Grand Junction canal, from which they easily get those cumbersome articles, it might be unprofitable or inconvenient for the railway to carry.

---

1 Brabazon Smyth Stafford, Chief Accountant at Wolverton.

The form of the building com[rising the factory is quadrangular; the offices of the factory clerks, draughtsmen, &c., being at the entrance; and the larger area in the middle filled with wheels and axles, most of them condemned as unfit or unworthy of trust for the public safety. On the right hand is the hospital for invalid and crippled locomotives. In this hospital I understand more sound cures are performed than any in London. Dr. Bury, the physician in chief, is esteemed a very clever man; and though not an Abernethy in conduct, his medicine is hard, and his treatment of the patients remarkably rough. he conceives there is much efficacy in mineral medicines, fiery ordeals, and hard lows, than in gentle laxatives and bland treatment.

The west side of the factory is chiefly occupied by the smithery: here are 24 forges, with slotting, planing, turning, boring, and every species of machinery necessary or useful for the making or repairing of the various parts of locomotives. The whole is driven with a high-pressure steam-engine, of 16 horse power; and to prevent the possibility of delay from accident to the engine, which would stop the whole establishment, they have two engines, one only, however, working at the same time. A division of labour is the necessary result of these arrangements, producing better articles, and at a cheaper rate to the Company. It is so managed that the forging and coarser and heavier work are carried on on the ground floors, and the finer work is done above. As nearly as I could judge, there appeared to be about 150 men employed in this side of the building alone.

It must be borne in mind that the whole of the Wolverton establishment is purely for the repairing of locomotives and their tenders. The carriage department is carried on at Euston-square. Both are repairing establishments: the Company manufacture neither engines nor carriages.

The south end of the Wolverton factory is chiefly occupied with compartments for putting the tires on wheels, and other purposes, furnaces for making and heating the tires, an oven for semi-coking the coal, in order to deprive it of its sulphur, which would prove injurious to the iron.

On the east is the tender hospital and establishment, model and other rooms; and on the left of the entrance into the factory stands what is significantly enough called their stable. Here are perhaps ten sound and able engines ready to take the place of any weak or crippled ones that may be forced to seek the benefit of rest and our Dr. Bury's assistance. two of them always have their steam up, and are ready to start at a moment's notice to give their assistance wherever it is wanted.

That which made the greatest impression on me in the Wolverton establishment was the systematic and methodical manner in which the business is conducted. In one part of the east end of the buildings is a series of large open drawers, numbered to correspond with the numbers of the engines; for on this railway the locomotives are distinguished by numbers instead of names. Here, as soon as an engine comes into the hospital, and is taken to pieces, every part of the small machinery is sent and deposited in its proper drawer; and should it not be wanted for a month or two, there it remains, without it being mixed or confused with any other, and is found when required without trouble or difficulty. each engine also has a separate account kept for it. It is debited for repairs and new materials, and credited with the old, or any which it returns to the stores. By this means, and as an account is also kept of what work the engines respectively perform, it is soon found out which kind of engine is the best, and the most desirable in future cases of new ones, to imitate or adopt; and as the engines, though in external appearance are nearly copies of each other, but made by different individuals, this debtor and creditor account keeps up a kind of internal competition, as conducive to the welfare of the Company as it is to the improvement of locomotive machines.

I have been the more particular in the description of this great establishment than I otherwise should, because it was the first of its kind for a Railway Company and has been the model for all others to imitate, and because it will hereafter save me the trouble of detailing those I may meet with other railways.

One point I have omitted to notice, which conduces very much to its efficiency; I allude to the centrical situation of the place between Birmingham and London. It is no matter much where the repairing establishment for carriages is, but the depôt for locomotives should evidently be about midway between the two grand termini, in order to afford the promptest aid in case of accident or casualty. Such is the situation of Wolverton.

The Census of 1841

The Census of 1841, while not the first taken nationwide, was the first where individual records were preserved. It provides names, approximate age and occupation and tells us whether or not the resident was born in the county.

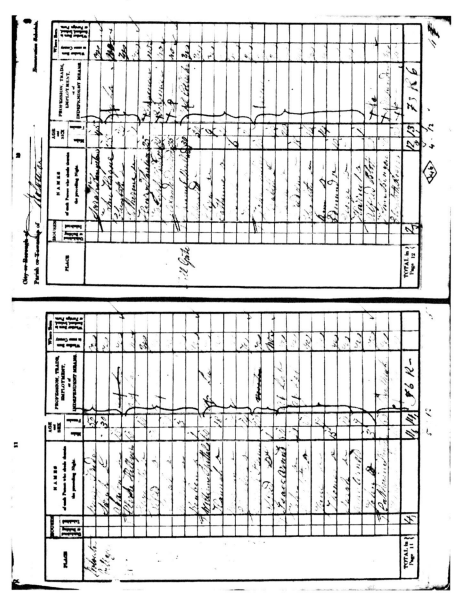

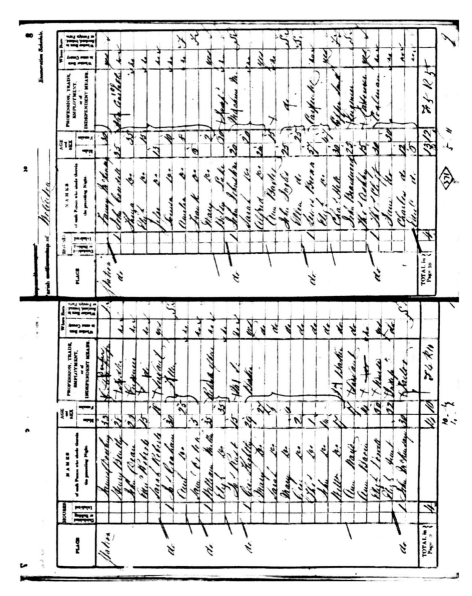

Already in 1841 "New" Wolverton had more residents than "Old" Wolverton. Two sample pages are reproduced here. The first is a page from the old village which shows mostly agricultural labourers living there, including a few lodgers, such as Alfred Blott the young station master. The second sample is from Wolverton Station, showing a page from Bury Street with several new shopkeepers.

The Directors and Chief Officers

LONDON AND BIRMINGHAM RAILWAY.

Office, Euston Square Terminus.

Chairman.—George Carr Glyn, Esq.

Deputy-Chairman.—Jos. Fdk. Ledsam, Esq.

Directors.James Brownell Boothby, Esq. Edmond Calvert, Esq. Thomas Cooke, Esq.Edward Cropper, Esq. Robert Garnett, Esq. Pascoe St. L. Grenfell, Esq. David Hodgson, Esq. Capt. C. R. Moorsom, R.N. John Lewis Prévost, Esq. Theodore W. Rathbone, Esq. C. S. Titmarshe, Esq. Thomas Smith, Esq. Thomas Tooke, Esq. Joseph Walker, Esq. Christopher Hird Jones, Esq. Thomas Young, Esq.

Consulting Engineer.—R. Stephenson, Esq.

Resident Engineer.—R. B. Dockray, Esq.

Superintendant.—H. P. Bruyères, Esq.

Superintendant of Locomotive Department. E. Bury, Esq.

Architect.—P. Hardwick, Esq., F.R.S.

Secretary.—R. Creed, Esq.

Solicitors, London. C C. Parker, Esq., G. Barker, Esq. Birmingham. S.Carter, Esq.

From Robinson's Railway Directory, 1841.

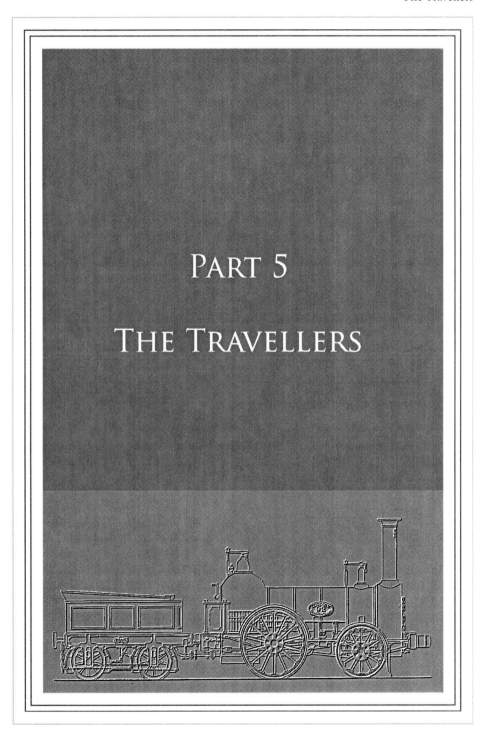

PART 5

THE TRAVELLERS

In this section

The new railway, the first to come down from the industrial north and the first to come to London, excited great interest. Travellers quickly multiplied and it was not long before travel guides (the most famous being Bradshaw) began to appear in print. The relevant sections of their printed books are reproduced here. It is interesting to note that they assumed that their readers would use the railway to reach Wolverton and explore the surrounding countryside from there. Here is a selection.

- James Drake,

a Birmingham printer and stationer was one of the first to enter this field.

- Thomas Roscoe (1791–1871)

was born in Liverpool, the son of a lawyer and MP. He was well educated and his work includes thetranslation of German, Italian and Spanish novelists into English. His travel writing was mainly devoted to Italy and Spain but he produced the occasional book on his native island, one of them being The London and Birmingham Railway, published in 1839.

- E.C. and W. Osborne

were a firm of Birmingham printers who, like Drake were swift to cash in on the new opportunities that rail travel created.

- Arthur Freeling

published a series of books in the early 1840s styled Freeling's Railway Companions. The books were aimed at the middle class leisure traveller. The books were advertised as " containg a complete description of everything worthy of attention on the line; of the gemtlemen's seats, villas, towns, villages, rivers, markets and fairs. An account of the churches, endowments, livings and patrons. List of races; and an account of the hunting and angling stations near the line."

PART 5: THE TRAVELLERS

Drake's London and Birmingham Railway

After rapidly sweeping through a cutting, we cross the London road by a stupendous iron bridge, which has a most noble appearance from below, and come to what was formerly known as the Denbigh Hall station.

Here, for several months after the first opening of the railway, the trains were accustomed to stop, and the traveller had to adopt the ancient methods of conveyance, for the performance of the next thirty-eight miles of his journey. To describe in all its serio-comic reality the scene which this now secluded spot was wont then to present, would require the pen of a Washington Irving. Luggage lost, tickets missing, coaches overfilled, and a thousand other disastrous occurrences, altogether formed a spectacle which we would defy the most sorrowful disciple of Heraciitus to view without a smile. All the busy multitudes, however, that so lately thronged this spot, and rendered it a scene of intense animation, have now vanished, like the fabric of Mirza's vision ; and as we rapidly sweep by, and look in vain for some tokens of animation, we are reminded of the feelings which travellers have had while sitting on the ruins of some ancient city. The building called Denbigh Hall, respecting which it is very probable our reader may have formed the same conception as ourselves, and imagined it to be the august mansion of some illustrious grandee, is nothing but a paltry public house, or " Tom and Jerry shop," as we heard an indignant fellow-traveller contemptuously style it, which has taken the liberty of assuming this magnificent appellation. Tradition ascribes the origin of the name to the ircumstance of Lord Denbigh having been compelled, to tarry here for a night, through an accident happening to his carriage ; and also informs us that his ordship left some property to his host in return for the kindness with which he had been entertained ; but whether this story is deserving of credit, or has merely been invented for the amusement of the visiters at this Denbigh Hall, we pretend not to say. After leaving this ci-divant station, and passing through a cutting three quarters of a mile in length, we perceive on the left the church of Loughton, and also that of Shenstone, which is a very good specimen of the Norman style of architecture. Close to the line on the right is the village of Bradwell, where was formerly a priory of Black Canons, founded in the reign of Stephen, and of which the abbey, transformed into a farm house, may still be seen standing on the left of the line. A short cutting, which is crossed by a bridge handsomely faced in a rustic style, brings us to Wolverton station.

WOLVERTON STATION.

Distance to London, 52½ — Birmingham, 59½ miles.

DISTANCES BY ROADS FROM THIS STATION TO THE FOLLOWING PLACES! —
Places W. of station.

Buckingham.	10½ miles.
Brackley	17½ miles
Stoney Stratford	2 miles

Places E. of Station.

Wolverton	1 mile. [2]
Newport Pagnell	4 miles.
Fenny Stratford	2 miles
Olney	9 miles

This being the central, and consequently the most important station between London and Birmingham, the buildings connected with it are on a scale of unparalleled magnificence. In addition to the locomotive engine house on the left, where these immense machines are manufactured, repaired, and kept in store, there is an extensive depot for goods on the right, and an area of several acres set apart for the reception of cattle. The style of architecture chiefly employed is the Doric; the beautiful simplicity of which harmonises well with the character of the buildings. But no useless ornament is employed : all is simple, grand, and imposing. Those passengers who wish to take some refreshment after a ride of fifty-two miles, have here ten minutes allowed them for that purpose. The town of Wolverton, hitherto unnoticed on the map of Great Britain, is now rapidly rising into importance ; houses are springing up on every side, streets are

---

2 The distance is right but the direction wrong.

being laid out, and a large and busy population is rapidly gathering ; whilst its fame as the birthplace of English fire steeds is spreading through the civilized world. Previously to the commencement of the railway, it contained only 417 inhabitants; but now, the railway company alone give direct employment to nearly a thousand hands.

This station will be found the most favourable for travellers proceeding to the towns of Stoney Stratford, Buckingham, Newport Pagnel, and Olney. The first of these places stands on the banks of the Ouse, one mile south-east of the station, and contains 1,700 inhabitants. It is celebrated in English history as having been the place where Eichard III., when Duke of Gloucester, seized Edward V. It has suffered greatly from accidental fires, 53 houses having been burned to the ground in 1736, and 113 in 1742. Prior to the introduction of waggons, it was a noted place of rendezvous for pack horses conveying goods to London, and the traffic through it is still very great.

Eight miles south-west of Stoney Stratford, is the ancient county town of Buckingham. Respecting the derivation of its name, etymologists differ widely ; but it appears most probable that the Saxon 'Bucca' which signifies a stag, lies at the root, since, in the early ages, the neighbourhood abounded with forests well stocked with deer. It is pleasantly situated on the river Ouse, which nearly encompasses the town, and is crossed by three stone bridges. The trade chiefly consists in the sorting of wool, the tanning of leather, and the manufacture of lace. The church stands on the site of an ancient baronial castle. It is a handsome structure, with a square embattled tower, and is internally elegantly fitted up in the Grecian style of architecture. Two miles west of Buckingham is Stowe

Park, the magnificent seat of the Duke of Buckingham and Chandos. The mansion was originally built by Sir Richard Temple, K.B., who died in 1697 ; it was enlarged by his son, Lord Cobham, and was brought to its present state of unrivalled magnificence by the late Marquis of Buckingham. The gardens or pleasure grounds of Stowe are more celebrated than even the mansion itself : they comprehend a space of more than 500 acres ; and contain a broad lake, a beautiful cascade, and a noble monument to Lord Cobham; together with a profusion of statues, temples, and every species of architectural adornment. A building in the flower gardens contains the mineralogical and geological collections of the Abbe

Haiiy, and an immense number of specimens in every branch of natural history collected by the Duke of Buckingham.

Newport Pagnell is a well built market town, lying six miles north-east of the station, and containing 3,385 inhabitants. The latter part of its name is derived from the family of Paganell, to whom the manor descended from the powerful baron, William Fitzansculf who held it at the time of the conquest.

The church stands on an eminence which affords a fine prospect of the surrounding country ; and in the churchyard may be seen the beautiful epitaph, written by Cowper, on Thomas Abbott Hamilton.

The other town which we mentioned as lying at a convenient distance from the Wolverton station, was that of Olney. This town lies ten miles north-east from the station, and, in common with the two last mentioned places, stands on the banks of the Ouse. It has a population of 2,418. The bridge over the Ouse is a handsome structure, consisting of five large arches, and two smaller ones. In the church, which is a large and ancient edifice, an unusually large number of celebrated literary personages have regularly officiated ; amongst whom we may notice Moses Browne, author of Piscatory Eclogues ; John Newton, the popular preacher and writer ; Thomas Scott, the celebrated biblical commentator ; and Henry Gauntlett, who wrote on the Apocalypse. Of all the great names, however, that are associated with Olney, there is none which recals so many pleasing remembrances as that of the poet Cowper. It was to this place that he retired to seclude himself from intercourse with a world, the rude gaze of which was alone sufficient to frighten his timid spirit ; and here, under the pastoral care of the Rev. John Newton, referred to above, he was in some measure relieved from that deep religious despondency into which he had fallen, and was enabled to form truer conceptions of that Divine system of religion which professes to be to all mankind glad tidings of great joy. Should our traveller be visiting Olney, we would sincerely recommend him to pay a visit to the house and garden of this amiable poet; and if he has ever dropped a tear on the grave of Byron's dog, in Newstead Abbey, perhaps he may not be unwilling to bestow the same tribute of sympathy on Cowper's hare, in his garden at Olney ; for, although Puss may not perhaps have been bewailed in elegiac strains quite so pathetic as those inscribed on Boatswain's tomb, yet her memory, also, is preserved in immortal verse, and future ages will hear of her innocent attempts to divert the melancholy of her sorrowful master.

WOLVERTON TO ROADE. Seven miles and a half.

Upon leaving Wolverton station, we behold directly before us the lofty steeple of Hanslope church, which, in point of conspicuousness, may almost vie with that of Harrow church. The delightful prospect which is now unfolded before us in every direction, includes Bradwell Wharf, Linford, and Mill Mead, on the right, and the village of Wolverton on the left. After crossing the Grand Junction Canal by a handsome iron bridge, and the Newport Pagnell and Stratford road by one of a more ordinary description, we arrive at the stupendous viaduct over the Ouse valley. This magnificent structure consists of six arches of sixty feet span, besides six smaller ones placed in the abutments ; and, to a spectator in the valley below, presents a most noble appearance. The view of the surrounding country, from the viaduct, is also exceedingly interesting. That on the right is thus beautifully described by a hand more graphic than ours : —

" Here Ouse, slow winding through a level plain
Of spacious meads with cattle sprinkled o'er,
Conducts the eye along his sinuous course
Delighted. There, fast rooted in their bank,
Stand, never overlooked, our favourite elms,
That screen the herdsman's solitary hut ;
While far beyond and overthwart the stream,
That, as with molten glass, inlays the vale,
The sloping land recedes into the clouds ;
Displaying on its varied side the grace
Of hedge-row beauties numberless, square tower,
Tall spire, from w liich the sound of cheerful bells
Just undulates upon the listening ear,
Groves, heaths, and smoking villages remote."

The Task.

One could almost imagine that the poet had written these lines while leaning on the parapet of the viaduct, and viewing the distant spire of

Haversham church, and the pretty cottages of Mead Mill. On the left the scenery is not less interesting. There also —

The Ouse, dividing the well-watered land,
Now glitters in the sun, and now retires
As bashful, yet impatient to be seen.

And not far distant is the stupendous embankment and cast iron viaduct by which the Grand Junction Canal is carried over the valley ; the towers of the two churches at Stoney Stratford rise above the viaduct ; Wolverton is seen among the rich foliage on the extreme left ; whilst the village of Cosgrove appears a little more in advance, and Castle Thorp in the distance. After the termination of the Wolverton embankment, we pass through a short cutting ; and then proceed along another embankment, through some finely wooded country, with a fertile valley on the left, and the village of Hanslope, with its lofty church spire, which now appears to the greatest advantage, on the right. Another cutting. a quarter of a mile in length, being passed, we discover on the left the villages of Stoke Bruern, Yardley Gobion, Pottersbury and Furtho, and also Whittlebury Forest, all of which are in the distance ; whilst, amongst the adjacent woodlands, the picturesque village of Grafton Regis, with the tower of its venerable church, can be distinctly perceived. This village is celebrated in history as having been the place where the clandestine marriage between Edward IV. and the widow of Sir John Gray, of Groby, was solemnized.

The London and Birmingham Railway

Thomas Roscoe picks up the journey from Denbigh Hall bridge, which , until the full line was opened on September 17th 1838, had been the southern terminus of the line.

Immediately after passing the Denbigh Hall Bridge we enter a long and deep cutting, extending about three quarters of a mile, and varying from thirty to forty-two feet in depth, from which 551,329 cubic yards of earth and stone were removed. This excavation is spanned by a lofty and elegantly proportioned bridge of three arches. At the termination of this defile the village of Loughton, with its massive tower rising from amidst a clump of fir trees, forms a beautiful object to the left of the line; and proceeding onwards we again enter another short excavation having three bridges thrown across the Railway. The train presently emerges from the shadows of this cutting, and an interesting and picturesque view bursts upon the sight, including the village of Bradwell, with its singular, old barn-like looking church. Some slight embankments and cuttings succeed, in the last of which is a lofty bridge of three arches, of a semicircular form, thrown across the cutting, which being built of blue limestone, with a rock rustic face and bold chamfered joints, has a remarkably picturesque appearance as the train passes beneath it, and, at length, we arrive at the Wolverton Station, one of the most important points on the line, distant from London fifty-two, and from Birmingham sixty miles, and about 200 feet above the level of the sea. The nearest town to Wolverton Station is Stoney Stratford, situate about two miles to the west, on the great Holyhead road.

The magnitude of the works at Wolverton is the wonder and admiration of all who travel along the Railway, and it will readily be perceived that the utility of having a great central station, on such a long line of road, was one of the first considerations; and it fortunately happens that the site is locally convenient as regards communication laterally by roads, and also by canal. It was therefore determined, that at this place a large manufactory should be erected, for the purpose of repairing engines and other machinery connected with the traffic; and, also, that a depot for the reception of goods and cattle should be provided; whilst accommo dation was secured by the erection of dwellings for the artificers ; it being contemplated that when at full work, nearly one thousand persons are employed at this station alone.

The large building seen to the left of the Railway, before arriving at the canal, is the Wolverton Locomotive Engine Station, which will be devoted wholly to the repairs, &c, of the engines and machinery; as the locomotives do not run through the whole length of the line, but change at this station, it is requisite that there should always be a stock in readiness to meet any demand. It was erected from the designs and under the superintendance of Mr. G. Aitcheson, architect of London, a gentleman who has for some time past devoted his attention to works of this description, and who has the appointment of architect to the Stations upon this line.

The building is of a quadrangular form, each side being 314 feet long; it is built of brick, with stone cornice and blockings, the style of architecture being Doric. But here there appears no extraneous ornament, all being in character with the objects for which it is designed; yet the extent of the building is so great, that it has a very imposing appearance. The entrance from the Railway is by an arch of a semicircular form, which leads to a large open area, entirely surrounded with buildings. On either side of the entrance are the erecting shops for engines; and around the court-yard are the engine and tender sheds, the joiners' shop, iron foundry, boiler yard, hooping furnaces, iron warehouse, smithy, turning shops, offices, stores, a steam-engine for giving motion to the machinery, and for pumping water into a large tank over the entrance gateway, to supply the locomotive engines; in short, every convenience that a large manufactory of this nature can require is provided. Between the building and the canal, a space of about two acres is set apart for a wharf, and store yard for timber, trucks, &c.; and on the eastern side of the Railway, opposite the engine station, is a space of several acres appropriated to the reception of goods and cattle ; it is bounded on the north and east by the Grand Junction Canal, (which here bends round towards the southward) and on the south by an occupation road, the Railway running on the western side. A layby has been already formed on the canal, and a large shed erected adjoining, for the purpose of receiving goods from the canal to be forwarded by Railway; this has been in operation since the entire opening of the line, and trains of goods now run regularly from hence to London. Waiting and refreshment rooms are erected close to the station for the accommodation of passengers, as this is the only place between London and Birmingham at which time is allowed to take refreshment. It is contemplated that on the Station, and the works connected with it, a sum of £100,000., has been expended.

The Railway next crosses the Grand Junction Canal, at an elevation of fourteen feet, by a neat iron bridge formed of flat ribs, the outer ones being pannelled; an iron railing extends over the whole length of the bridge and retaining walls, which gives it a light appearance. The bridge is extended on the eastern side of the Railway, where a newly formed road crosses it, leading to the station and offices. Immediately after passing which we enter on the great Wolverton Embankment, formed across the valley of the Ouse. This embankment is the largest on the line, being one mile and a half long, ave raging 48 feet in height, and containing 500,000 cubic yards; nearly in the centre of it stands the Wolverton Viaduct, beneath which flow the diverted

channel of the rivers Ouse and Tow. The viaduct is built of brick of a peculiarly fine colour, and the stupendousness and grandeur of its propor tions are best observed from the meadows where Mr. Dodgson's view was taken. It consists of six elliptical arches of sixty feet span each, rising twenty feet; the height to the soffit being forty-six feet. At either end are two massive pilasters, with stone cornice and blockings, and beyond are three smaller arches, which pierce the retaining walls, built n the slope of the embankment; the cornice is continued throughout the whole length of the viaduct, and is surmounted by a parapet wall, the top of which is fifty-seven feet from the level of the ground; the total length of the viaduct is 660 feet, or one eighth of a mile, and the cost of its erection was £28,000.

In contemplating this magnificent structure the eye is forcibly struck with the beauty of the design, and elegance of its proportions, whilst the masterly execution of the work reflects great credit on those who were engaged in its erection.

In the formation of the embankment at Wolverton, great difficulties were encountered. On the north side of the Viaduct the material is composed of blue clay, lias limestone, gravel, and sand. This part of the embankment stood very well, except in one place where it slipped, not on account of its being composed of bad material, but from the ground itself actually yielding when the weight of the embankment came on it. The length of the embankment being one mile and twenty-eight chains (deducting the viaduct), and the height of a great part of it forty-eight feet, some accidents were to be expected, especially in bad weather; but no one could have imagined what would take place on the south side of the viaduct. Here the material, at the commencement, was composed of sand, gravel, and blue clay: this stood very well; but when the workmen went deeper into the cutting, they excavated some black, soapy clay; this was tipped on to a turf bottom, and the weather being also very unfavourable, although every care was taken to mix dry stuff with the wet material, yet there occurred one of the worst, if not the worst slip, along the whole line. Earth was tipped in for days and days, and not the slightest progress was made; as fast, in fact, as it was tipped in at the top it kept bulging out at the bottom, till it had run out from 160 to 170 feet from the top of the embankment; and at last a temporary wooden bridge was formed, and, by wagoning the earth over this, the embankment between the slip and the viaduct was formed, by first digging a trench, five feet deep, and nearly the whole width of the embankment, and forming a mound on each side to prevent it from giving way.

In fine Summer weather the bridge was removed, and that part of the embankment, where the slip had been, was filled up; but away it went again, just as it did before, and the yawning gulf appeared to be insatiable.

It was only after incredible labour and patience that it was conquered, and this was done by harrowing as much earth to the outer part of the slip as would balance the weight on the top.

There seemed to be no end to the vagaries of this embankment. There was a portion of alum shale in it, which contained sulphuret of iron ; this becoming decomposed, spontaneous combustion ensued, and one fine morning there was the novel sight of a fifty feet embankment on fire, sleepers and all, to the great surprise of the beholders. The inhabitants of the neighbouring villages turned out, of course, in no small amazement on the occasion; and various were the contending opinions as to the why and the wherefore. Some said—" The Company were hard up for cash, and were going to melt some of the rails;" others, " that it was a visitation of Providence, like the tower of Babel." At last one village Solon settled the point—" Dang it," said he, " they can't make this here Railway arter all; and they've set it o'fire to cheat their creditors."

Osborne's London and Birmingham Railway Guide

On the right is the village and church of BRADWELL, and close to the line is a quarry, from which materials have been taken to construct the neighbouring embankments; in this the blue clay may be seen resting on the oolite limestone, which appears of a yellowish colour, having been exposed to the weather, but is blue when first broken; it is of a very hard texture, and has been used for building the neighbouring bridge. Just after this, the train stops at the grand central depot of the company,

WOLVERTON STATION.

From London 52 ½ miles.—From Birmingham 59 ½ miles.

This Station is an extensive establishment, and will probably give rise to the formation of a new town; the engine house, which is 112 yards square, is on the left, and the booking offices on the right side; upwards of fifty cottages belonging to the servants of the company have already sprung into existence. A stall at which confectionary is sold, is permitted at this station. Just beyond the station, the road from Newport Pagnell to Stony Stratford is crossed by a bridge, from which, on the left, towards Stony Stratford, WOLVERTON CHURCH is observable amidst the trees, about half a mile off. STONY STRATFORD is a market town in Bucks, two miles on the left of the station; it consists principally of one street, situated on the parliamentary road from London to Birmingham, Holyhead, Chester, and Liverpool, and contains about 1700 inhabitants. The name is supposed to be derived from a ford of the Ouse, anciently situated about here. Owing to the great thoroughfare, there is considerable traffic carried on; but the only manufacture is that of bone lace: the market is held on Friday, and there are several fairs in the year.

BUCKINGHAM, ten miles and a half on the left, is a borough and market town of considerable antiquity; the present name is Saxon, buck, signifying a stag, ing a meadow, and ham a village; the neighbourhood was in ancient times celebrated for its forests and deer. This town suffered much from the incursions of the Danes, and the remains of walls erected on the sides of the Ouse by Edward the Elder, to protect the inhabitants from these depredations, are still distinguishable. In 1724, a dreadful fire took place in the town, by which several streets were burnt down: many of these houses are not yet rebuilt.

The town stands on a peninsula formed by the river Ouse; it nearly encircling the town, constituted in former days a natural moat; there are several bridges, two of which are of very ancient date. The lace trade formerly employed the inhabitants to a considerable extent; but this has been much limited by the improvements in machinery that have taken place at Nottingham. There is a general market held on Saturday, and one exclusively for calves on Monday. Fairs, chiefly for the sale of cattle, are held on old New Year's Day, the last Monday in January, March the 7th, the second Monday in April, May the 6th, Whit Thursday, July 10th, Sept. 4th, Oct. 2nd, Saturday after old Michaelmas Day, Nov. 8th, and Dec. 13th. Buckingham gives the title of Duke to the family of Grenville.

At a few miles distance, northward from the town, beyond Stowe, is TINGEWICK, the seat of the Duke of Buckingham and Marquis of Chandos, celebrated for its costly splendour and magnificence: at a distance the visitor is led to anticipate much, from the number of towers, columns, obelisks, &c, which rise above the foliage ; and on arriving, there is so much of grandeur, beauty, and taste exhibited, that the description would be adequate to fill volumes. The grounds are ornamented by temples, pavilions, arches, grottos, gardens, statues, and groves; the mansion, which is situated on a sloping lawn, faces the south, its front consisting of a centre connected by colonnades to two pavillion wings; a projecting pediment, supported by Corinthian columns, is at the entrance; and from this down to the lawn is a flight of magnificent steps, at the bottom of which are two large stone lions. The interior vies with the exterior in all that is superb: the saloon, the halls, the drawing-rooms, music-room, library, state galleries, state bed-chambers, state closets, &c, are all of a character equal to the rest of this princely establishment.

Four and a half miles to the right is the town of NEWPORT PAGNELL, one of the largest in Buckinghamshire ; it is situated on a slight eminence, and the main street is well built, but deficient with respect to a good pavement and lighting.   The river Levet runs through the town, and supplies it with water. The addition of the name Pagnell results from a distinguished family of baronial chiefs, who anciently lorded it over the manor. The manufacture of bone lace used to give considerable occupation to the inhabitants of this town; but the competition of Nottingham machinery has caused the trade to decline: the manufacture of paper gives a good deal of employment. Annual races were re-established here, after an intermission of forty years, in 1827, and are now regularly held every August. The market for bone lace is on Wednesday, and a common market on Saturday: six fairs are held here in the course of the year. The reform bill has made Newport Pagnell a polling place for the county of Bucks.

Just after the train has crossed the bridge over the road, we arrive upon the great viaduct over the Ouse, which consists of six large central arches, and three small terminal ones at each side, making in all twelve arches.

To the left, the canal carried over the Ouse on a viaduct, and across the valley on a high embankment, may now and then be seen; and in the intermediate fields, which are flat, the Ouse winds along in various directions, fringed in part with aged willows. A little onward, on the same side, the village of COSGROVE Stands upon a hill, the church visible above the trees. On the right side, the tall spire of HANSLOPE CHURCH, situated on a high eminence, is particularly conspicuous in the onward distance.

The valley of the Ouse being crossed by a high embankment, we enter the Castlethorp Cutting through the oolitic limestone, and then find ourselves upon another embankment over a wide valley, with the spire of GRAFTON visible on the hill, towards the left, and HANSLOPE CHURCH standing conspicuous to the right; the valley is of a luxuriant nature, consisting of pastures filled with cattle, rich corn fields, and here and there a farm house, with its orchard and other rural appendages. HANSLOPE is principally celebrated for its high spire, which is observable for miles throughout the surrounding country; this steeple was thrown down by lightning in 1804, on a Sunday evening in June; fortunately the church was empty, and no lives were lost: the spire has been rebuilt in the original form.

The London and Birmingham Railway Companion

We now pass under Bradwell Abbey Bridge; and, a little past the 51 ½ post, we rapidly pass through a short excavation, and arrive at the WOLVERTON STATION.

	miles	1st class. 4 inside by day, or 6 inside by night.	1st class 6 inside by day	2nd class closed by night	2nd. Class open by day

F r o m L o n d o n, rather more than	52 ¼	15s. 6d.	14s.	11s. 6d.	9s. 6d.
F r o m Birmingham, rather less than	60	17s. 6d.	16s.	13s. 6d.	10s. 6d.

From this station[3] Fenny Stratford is 6 miles; Newport Pagnel, 4; Olney, 9 miles eastward, or to the right of the line; Stony Stratford, 2; Buckingham, 10 miles to the eastward, or left of the line.

NEWPORT PAGNEL is a market town and parish, in the hundred of Newport, county of Bucks; it is 4 miles west of the line, and 51 N.W. from London; pop. 3,385; An. As. Val. £9,208; market, Saturday; fairs, Feb. 22d, April 22d, June 22d, August 28th, October 22d; and for cattle, December 22d. This town, one of the largest in the county, is situated on a gentle eminence. In the lime of the Civil Wars it was held for the King by Prince Rupert, until 1643, when the Earl of Essex took possession of it for the Parliament. Sir Samuel Luke, supposed to be the Hudibras of Butler, was governor in 1645. The principal manufacture of the town is lace-making, and it is supposed to make more than all the other parts of -England put together; the lace-market is held on Wednesday. Water is supplied to the town by wells, and a branch of the Grand Junction Canal, which comes up to the town, enables the inhabitants to obtain their coal from Staffordshire.

The church, dedicated to St. Peter and St. Paul, is an ancient and spacious structure, with a square tower, situated upon a considerable eminence, which commands delightful views of the surrounding country; the living is a dis-vicarage, in the archdeaconry of Buckingham, and diocese of Lincoln; K. B. £10; it is in the patronage of the Crown, and is endowed with £400 by royal bounty and private benefaction. Here are places of worship for Baptists, Independents, Wesleyans, and Presbyterians; an endowed school for twenty girls; a national, and a Lancasterian school, supported by voluntary contributions; Queen Anne's Hospital, endowed

---

[3] From arrangements which have been entered into by the Company with respect to the coaching department, the Leighton Station will be the most convenient for persons going to Fenny Stratford

for six poor men and women; and a close at North Crawley, for the widow of any vicar of the parish; when there is no such person living, the rental is appropriated to apprenticing poor children. Cowper the poet lived near here for some time, and died in the year 1800.

OLNEY is a market town and parish, in the hundred of Newport, county of Bucks; is 9 miles from the railroad, and 55 N.W. by N. of London; pop. 2,344; An. As. Val. £6,489; market, Monday; fairs, Easter Monday, June 29th, and October 21st, for cattle. The town is situated on the north bank of the river Ouse, over which is a bridge of four large, and several small arches, forming a sort of an embankment across the adjoining lands, which are frequently overflowed; there is an especial fund appropriated to keeping it in repair. The principal manufacture is that of bone lace; worsted hose and silk weaving have also for some time been carried on here to a considerable extent.

The church, dedicated to St. Peter and St. Paul, is an ancient and spacious edifice, in the English style of architecture, with a tower and lofty spire; the living is a vicarage, in the archdeaconry of Buckingham, and diocese of Lincoln; K. B. £13 6s. 8d., endowed with £700 by parliamentary grant and private benefaction; P. R. £100; patron, the Earl of Dartmouth. Here are also places of worship for various denominations of Dissenters, and the Society of Friends; alms-houses for twelve poor women; and national and Lancasterian schools, supported by subscription. The celebrated Newton, the friend of Cowper, was curate of this church. Cowper's residence is about a mile from the town, and, with his garden, and favourite seat, is still visited by the curious.

FENNY STRATFORD is a market town and chapelry, partly In the parish of BLETCHLEY, and partly in the parish of SIMPSON, in the hundred of Newport, county of Bucks; pop. 635; An. As. Val. £3,291; it is l ½ mile east of the line, and 45 miles N.W. from London. Market on Monday; fairs, April 19th, July 18th, .October 16th, and November 28th, principally for cattle. The principal manufacture is lace-making; the town chiefly depends on its local trade and the visits of travellers. The Grand Junction Canal passes near the bottom of the town. It takes its distinguishing prefix from the nature of the neighbouring lands; it is situated upon a rising hill. The town suffered much from the plague, in 1665, in consequence of which the inns were shut up, and the high-road turned another way; this ruined the market, which, it is said, has never recovered its former activity. The chapel, dedicated to St. Martin, has no particular architectural pretensions; the living is a perpetual curacy, in the archdeaconry of Buckingham, and diocese of Lincoln; P. R. £105, endowed with £1,000 by private benefaction and royal bounty; patron, John Willis, Esq. (1829.) Here also are places of worship for Baptists and Methodists; also a national school.

STONY STRATFORD—a market town and parish, in the hundred of Newport, county of Bucks, C miles west of the railway, and 52 from London; pop. 1,619; An. As. Val. £2,088. Market on Friday; fairs, August 2nd, Friday after October 10th, and November 12th; the first and last are for cattle, the other principally for the hiring of servants. The original name of this town, Lactodorum, signifies, a river forded by stones; the present name is no doubt derived from it, as well as the fact of access -being had to the town by a stone bridge across the river Ouse. It was near here that Edward the Elder stationed his army while he fortified Towcester. Stratford is situated on the bank of the river Ouse, upon the old Roman Watling Street. It has a very considerable trade in corn, and a large quantity of lace is also manufactured by the inhabitants. The Grand Junction Canal passes through the town.

The church, dedicated to St. Giles, was rebuilt in 1776, except the tower. The united livings of St. Mary Magdalene and St. Giles's now form a perpetual curacy, in the archdeaconry of Buckingham, and diocese of Lincoln ; it is endowed with £800 by royal bounty and private benefaction, and also with an annuity of jG8. Here are places of worship for various denominations of Dissenters; two Sunday schools, with an endowment for ten boys; a national school, supported by voluntary subscriptions; and a fund of J670 for apprenticing children of the town or parish.

BUCKINGHAM—a borough, county, and market town and parish, in the hundred and county of Buckinghamshire, 10 miles from the railway, and 66 N.W. by W. of London; pop. 3,610; An. As. Val. £10,660. Market, Saturday; fairs, January 12th, and the last Monday in January, March 7th, May 6th, Whit Thursday, July 10th, September 4th, and October 2nd, principally for cattle and sheep; and a statute fair on the Saturday after old Michaelmas-day.

This place is of very remote antiquity; by some it is said to have derived its name from the extensive beech-woods which were formerly in its neighbourhood ; while others assert that it was adopted from the large quantity of deer which inhabited those woods; but, as the name was originally written Bockingham, others deduce it from the Saxon *Bocking*, a chartered land, and ham, which signifies a home — this appears a natural derivation. It has, at various times, been the scene of interesting historical occurrences. Here the Romans, under Aulus Plautius, defeated the Britons under Caractacus; it was fortified in 918, by Edward the Elder, to defend it from the Danes, who, however, took the place a century afterwards ; and it was repeatedly occupied by the belligerents during the Civil War between Charles I. and the Parliament.

The town is pleasantly situated on the side and bottom of a hill, upon the banks of the Ouse, over which are three bridges of stone. The principal occupation of the inhabitants is, the making of lace, tanning, wool-sorting, and paper-making, Cthere being several mills upon the river;) agricultural pursuits occupy the population not employed in the town. The Grand Junction Canal gives it the benefit of an extensive water communication.

The church, dedicated to St. Peter and St. Paul, is the principal ornament of the town, and is delightfully situated, upon an artificial mount, formerly occupied by the castle. It is a stately fabric, with a highly ornamented square embattled tower, surmounted by a finely proportioned spire; it is built of free-stone, on the model of Portland chapel, in London. The interior is peculiarly handsome, and the altar is embellished with a copy of Raphael's Transfiguration, presented by the Marquis of Buckingham, whose uncle, Earl Temple, contributed largely to the erection of the church; the living is a dis-vicarage, and a peculiar of the Dean and Chapter of Lincoln; K. B. £22; patron, the Duke of Buckingham. There are places of worship for Baptists, Independents, Presbyterians, and the Society of Friends; an endowed free grammar school; the green-coat school, endowed for the education of 26 boys; a national school, supported by voluntary contributions, in which are educated 200 boys, and 100 girls (a handsome and commodious building has been erected for their occupation); and almshouses for six aged women, founded by Queen Elizabeth.

This town sent members to Parliament from the time of Henry VIII.; the Reform Bill has extended the franchise to £10 householders; two members are returned for this borough. It gives the title of Duke to the Temple family. Two miles beyond the town, is Stowe, the magnificent seat of the Duke of Buckingham; its site may be distinguished by the appearance it presents of a vast wood. This is the most magnificent seat in the county; it is a residence worthy of a king, and far surpassing any of our royal palaces, if we except Windsor Castle; the edifice is situated in noble grounds; its principal front is 916 feet long; but, as an architectural description would take up more space than we can afford, we will assure our readers, if they should ever pass within ten miles of Buckingham, and have leisure to view this mansion and grounds, that they will consider a day thus spent one of the most agreeable and instructive of their existence. The interior of the edifice is equal to its exterior; taste is the presiding deity, and each department exhibits its influence. A library of 10,000 volumes offers further proof, if any were necessary, that a superior mind has governed every arrangement.

Before resuming our journey, we must remark, that Wolverton is the grand central station, at which are workshops, artisans, &c, and every convenience, to repair accidents, or to obtain any requisite which the trains may require: the works are well worth inspection, if the traveller has time. At this station the Grand Junction Canal is crossed by a massive iron bridge; the road to the right is for The convenience of passengers going to, or coming from, the railway. The village of Wolverton is about one mile to the left; it contains a population of about 350 persons, chiefly rural. The church is dedicated to the Holy Trinity: the living is a discharged vicarage; K. B. £10 3s. 9d.; in the archdeaconry of Buckingham, and diocese of Lincoln. Near this village is Wolverton great ground, a favourite meeting of the Duke of Grafton's hounds. The country, for some distance on each side of the railway, forms a part of the Radcliffe estate, left by the will of Dr. Radcliffe, as an endowment for the Radcliffe Library at Oxford. We proceed along this gigantic embankment, which is a mile and a half long, 50 feet high, and contains nearly eight hundred thousand cubic yards of earth. From such a height the view is, of course, extensive. We shall notice various objects as we arrive at them; in the mean time, we may observe, that the high steeple which is seen some miles in front, is a well-known landmark, being a portion of Hanslope church. A little past the 53 post, we cross the rivers Ouse and Towe by a stupendous viaduct, by which we pass from the parish of Wolverton to that of Haversham. This work is 210 yards in length, is 54 feet high from the level of the water to the top of the parapet; it consists of six centre arches of 60 feet span, and six smaller ones, which are in its abutments; the piers are ten feet in thickness. Here, to the left, a green mound may be observed, which is connected by a small bridge; it is about half a mile distant on the left: this mound is the channel of the Grand Junction Canal; it is cased with iron, and the object which appears like a bridge, is, in fact, a portion of this iron trough; under it is a field-road. Stony Stratford may be distinguished by its tower at about two miles distant upon the left hand. We may mention (en passant) that Richard III., when Dule of Gloucester, there arrested the Earl of Rivers, seized the person of the young prince Edward V., at an inn in the town, and, in his presence, also arrested Lord Richard Grey and Sir Thos. Vaughan. Prince Edward accompanied the Protector to London ; the other prisoners were conducted to Pomfret, where they were afterwards beheaded, under the direction of that contemptible tool of a blood-thirsty tyrant, Sir Richard Ratcliffe. Cosgrove church may be seen a little further on; the hamlet is situated upon a wooded declivity: Pop. 624; An. As. Val. £3662. The church is dedicated to St. Peter; the living is a rectory, in the archdeaconry of Northampton, and diocese of Peterborough; K. B.414 lis.; patron, Major Mansell. Cosgrove Priory, now a private residence, is close to the canal-bridge. To the left, beyond Cosgrove, is Whittlebury forest; it extends for about 11 miles along the southern border of the county; it comprises about 6,000 acres, and contains within its shades about 1,800 head of deer. The Duke of Grafton is Lord Warden of the forest, which is under his superintendence, assisted by a deputy warden, two verderers, a woodward, a purlieu ranger, five keepers, and six page-keepers.

We now enter a cutting through rubbly sand and limestone; in the latter were very strong springs of water. A little further on is the village of Castlethorp; its cottages may be seen from the railway, to which their gardens extend. Near the church was the ancient castle of the Barony of Hanslope; it was a fortress of considerable strength and extent, as the remains testify. . It was held against Henry the Third by its owner, Baron William Manduit, and was taken from him and razed to the ground by

Fulke de Brent. Near to it is a favourite meeting of the Grafton hounds. About here the excavation is interrupted by a small embankment of about 100 yards in length, from which we again see the tower and graceful spire of Hanslope church, at about a mile distant. Hanslope is a parish in the hundred of Newport, and county of Bucks; population, 1623. An. As.Val. £6,662. Lace-making employs many of its inhabitants. The church, dedicated to St. James, is a very ancient edifice; it is in the early Gothic style of architecture, and had formerly an octagonal, fluted spire, 200 feet high, which was destroyed by lightning in 1804 ; this was replaced by the present erection, so " remarkable for its graceful proportions; it is about 190 feet in height. The living is a rectory, in the archdeaconry of Bucks, and diocese of Lincoln; K. B. £48; C. V. £16; P. R. £106; patron, the Corporation of Lincoln.

From this embankment we have the first sight of the church of Grafton Regis; its low, massive tower rises just above the trees. We shall notice the hamlet when we arrive opposite it, which will be near the 56 ¾ mile post, just before entering an excavation.

Grafton Regis is seen on our left, perhaps a mile distant—the church tower points out its site: it is a parish in the hundred of Clely ; population 241, mostly employed in lace-making. The church is dedicated to St. Mary; the living is a discharged rectory, in the archdeaconry of Northampton, and diocese of Peterborough; K. B. £9 9s. 4{d. patron, the Lord Chancellor. At this place Edward the Fourth was married to Elizabeth, the beautiful widow of Sir John Gray, of Groby; one of those extraordinary instances of the power of love over interest, policy, and ambition, which proves that the romance of the imagination cannot surpass that of real life; many other historical events are con-i nected with this place, of which our space prevents notice. Grafton gives the title of Duke to the Fitzroy family. Close by this village, upon a wooded eminence, about two miles to the right, is Stoke Park, the residence of Wentworth Vernon, Esq. This elegant mansion was built about the middle of the seventeenth century, and took six years in completing. During the progress of the work, Charles I. and his Queen honoured the estate by partaking of a sumptuous banquet, at the invitation of the proprietor, Francis Orane, .Esq., upon whom Charles had conferred the estate. The mansion consists of a centre and wings, connected by corridors, and it is situated in a well-wooded park.

PART 6

LIFE IN THE NEW TOWN

In this section

- ## Hugh Stowell Brown: Notes on my Life

  was born on the Isle of Man. He came to Wolverton as a 16 year old and left three years later once he had discovered his mission in life, which was to become a Baptist minister. He was, in his day, a celebrated preacher, and a statue was erected to him in Liverpool. In late life he wrote an autobiography, *Notes on my Life*, which includes an account of his years in Wolverton. It was published in 1888.

- ## Building a Community

  Board minutes from the london and Birmingham Railway Board and the Radcliffe Trust about the issues of establishing schools and the church. Both organizations took their social responsibilities seriously. there was less government in those days.

  Letters from some of the protagonists on these matters.

  A newspaper report from The Times about the consecration of the church of St. george the Martyr.

- ## A report to the Radcliffe trustees by Edwin Driver

  A report concerning the L & BR Board;s need for more land and some of the social tensions between the new residents and the farmers.

- ## A report on health issues arising from the water supply

  A paper presented to the Royal Pharmaceutical Society about some diseases traced to Wolverton's well water.Salary Registers.

- ## Matters of Scandal and Concern

  Two stories about Victorian morality emerge from the minute books of the LNWR - one about the Station Master's elopement with an unnamed young lady and the other about a liaison between two schoolteachers.

  An article from The Bucks Herald, August 2nd. 1851 about a child murder.

- ## The Comedians

  The census of 1851 reveals a company of "Comedians" staying at the Radcliffe Arms and the New Inn. Wolverton people were not deprived of entertainment.

92

Part 6: Life in the New Town

Hugh Stowell Brown - extracts from *Notes on My Life*

Chapter VIII I Go To Wolverton.

I HAVE sometimes wished that I had continued on the Ordnance Survey, but I had made up my mind to become a mechanical engineer. Mr., afterwards Dr. Carpenter, of St. Barnabas, Douglas, was acquainted with one of the directors of the London and Birmingham Railway, and through Carpenter I obtained permission to go to the Company's works at Wolverton, whither I accordingly went in the month of August, after little more than a month of Ordnance Survey.

And here begins a new chapter of my poor history. The London and Birmingham was then the pride of English railways. It was indeed the best line of railway in the world, and I don't know a better even now. It was the greatest of Robert Stephenson's achievements. He had tried his 'prentice hand in the construction of the Liverpool and Manchester, and this larger work was his masterpiece. To this day on going from Liverpool to London, after going over the Trent Valley, the traveller, as he proceeds from Rugby to the metropolis, feels that he is on a firmer, smoother, better road. He is then on the old London and Birmingham. It had been opened throughout about two years when I entered the Company's service, Wolverton station, fifty miles from London and sixty from Birmingham, was regarded as a halfway house, and every passenger-train stopped there ten minutes for refreshments. I believe the Company's original intention was to take the line by Northampton, and to have their engine-works there, but the wiseacres of that town fought against the Bill, and so far succeeded in cutting their own throats as to divert the line, not allowing it to come within five miles of them. It was a grand mistake; they shut themselves out of the world. When they wanted to go to London or Birmingham they had to jog along those five miles to Blisworth station in omnibuses. Had the engine-works been built at Northampton, the town would have added to its trade perhaps £2000 a week, besides all the profit of building works and houses. For many years Northampton had to be satisfied with a branch line, and now (1879) after more than forty years the original design is in part being carried out by means of a loop from Roade to Rugby, whereby the long tunnel at Kilsby will be escaped. Had the Northampton people been wise that tunnel need not have been made.

The engine-works were then established at Wolverton. They consisted at first of a large, square building enclosing a quadrangular yard, and ordinary smith's shops, turning and fitting shops, erecting shops, pattern shop, iron foundry and brass foundry, with an engine-shed, and the number of men employed was about five hundred When I went to Wolverton there were not more than about twenty houses for the workmen. The majority of the men lodged in the villages round about—Old Wolverton, Cosgrove, Castlethorpe, Haulope, Bradwell, Calverton, and in fire towns of Stony Stratford to the west and Newport Pagnell to the cast. Wolverton is in Buckingharnshire, on the border of Northamptonshire, and in the valley of the lazy Ouse, which creeps through the railway, half a mile north of the station. Like the parishes which surround it, its occupation is wholly agricultural, with the addition of lace-making, carried on in the cottages by the women and children. The country is prettily adorned with many goodly trees, and many quaint and picturesque old churches; Haulope with its noble spire being particularly fine. Beneath the trees is embowered the little church of Cosgrove, where lie the remains of Dean Mansell. But more interesting is Olney, fragrant with the memories of Cowper and John Newton. It is also a pleasant walk to Wootton Abbey on one hand, and to the stately Stowe near Buckingham on the other. The geological structure of the district is the well-known oolite, so common in Northamptonshire and neighbouring counties. The two towns, Stony Stratford and Newport Pagnell, were, as they still are, very dull, dead-alive places. They had just been shorn of their glories. Each, situated on one of the great roads from London to the north, had seen many coaches pass through it every day. The railway had superseded every one of them. No longer was heard the guard's horn, no longer seen the well-appointed equipages, each with its four fine horses, and its proud driver with a bunch of flowers in his button-hole. Looking along the Stratford High Street, one saw on either side many sign-boards swinging in the wind—' The Cock,' 'The Bull,' 'The Cross Keys,' and at Old Stratford 'The Saracen's Head.' The number of inns, large and small, was numerous, each being patronized by its own coach or coaches. But now, small indeed was the custom at these once busy hostelries; and the grass grew long in Stratford Market-place, and we railway folk were looked upon with much disfavour. We had ruined the trade of the town; but yet I should think that after all Stratford lost little by the change. Most of the wages paid at Wolverton came into the hands of the Stratford shopkeepers, and not less than £100 was spent in the Stratford publics on every Saturday night by the 'station - men,' as they were called. And not a few of the 'station-men ' also, when they saw that the daughters of Stratford were fair, took them wives who fared much better than they were likely to have done but for these strangers. Still the talk of the townspeople was full of sad references to the good old coaching days.

Chapter IX Railways in 1840.

IN the summer of 1840, when I went to Wolverton, the traffic on the London and Birmingham differed greatly from that which we witness now after a lapse of thirty-nine years. It should be borne in mind that this line had all the traffic between London and the north. There was no Great Northern Company for the trade between London and Scotland; the Great Western had not gone to Birmingham; the Midland poured all its London traffic into the London and Birmingham at Rugby. The only connection between the metropolis and Northampton, Aylesbury, Coventry, Birmingham, Wolverhampton, Shrewsbury, Stafford, Derby, Leicester, Nottingham, Sheffield, Manchester, Liverpool, Chester, Holyhead, Leeds, Preston, Huddersfield, Bolton, Bury, Halifax, Bradford, Newcastle, Edinburgh and Glasgow, was by that one line from London to Birmingham. In fact, if you exclude the eastern counties, you may say that the whole of Great Britain north of London, together with the larger part of Ireland, depended for its traffic with the metropolis upon the one line of railway of which Wolverton was the centre. The Grand Junction at Birmingham, and the Midland at Rugby, brought goods and passengers from the north; coaches from many places ran to the various stations on the line, which was a gigantic monopoly, its only competitor in goods being the Grand Junction Canal; in passenger traffic it had no competitor at all. And yet to what did its traffic, its passenger traffic, amount? The Company ran per diem nine through trains each way, and two others, one between London and Wolverton, the other between London and Aylesbury, the branch from Cheddington to that town being the first and in 1840 the only branch in existence. And the trains were very light. There were, as far as I can recollect, few trains of more than ten carriages, each containing three compartments. They were little more than half the size of the carriages now in use. Those small trains and small carriages sufficed for all the passenger traffic of the vast district above defined. And now! Well, now, the Midland has withdrawn its share and runs to St. Pancras, taking to a great extent the midland county passengers, competing with the London and North-Western Railway for the traffic with Manchester and Liverpool, and all Scotland, and connecting with Northampton, and much more of the old London and Birmingham ground. The Great Northern, another formidable rival, has a large share of the northern traffic. The Great Northern bids against the old line for the trade with Birmingham, Leamington, Wednesbury, Wolverhampton, Shrewsbury, Chester. Thus the district once entirely monopolized by the London and Birmingham for London is now shared by three other companies. And yet how stands the case? In stead of nine through trains per day, there are thirteen, together with many more which run to and from Rugby, taking in Trent Valley for the north, and many more of a local character; and taking into account the number and size of the carriages, the passenger traffic at this day on the old London and Birmingham must be more than fourfold what it was when it had not a single rival to compete with.

It is interesting to notice the difference in speed. The fastest train, and there was only one such per day, was five hours on the road between London and Birmingham, and now in five hours we go nearly twice the distance—i. a. from London to Liverpool. The greatest distance run without a stoppage was from London to Tring, 31 miles, and now we run from Willesden to Rugby, about 76 miles, without stopping. The fastest train stopped four times between London and Birmingham; it now stops only once at Rugby. There were sixteen stations between the termini; there are now twenty-four. The fares were high: first-class, London to Birmingham, 32s. ; second, 25s.; but there was a second-class which was open to the weather, and the charge by that was 20s. And now to Liverpool, the first-class is 29s. 6d., and the second, as good as the first was then, 21s. 9d. In fact, the present fares are little more than half what they were in 1840. There was no third-class, and working people could not afford second-class fares; so they went on foot. Our workmen at Wolverton came and went on tramp from Lancashire and from London, and when discharged walked back again, or elsewhere, in search of work. Almost every day when we turned out at the dinnertime we found some half-dozen tramps, smiths, fitters, turners, boiler-makers, sitting under the wall of the shop in very shabby clothes, with blistered and bleeding feet, and to show them hospitality by taking them to dinner, was one of the prime duties that devolved upon us. The railway was not of the slightest advantage to workmen who had to travel. It probably was rather a disadvantage. In former days a lift on the coach or on the carrier's waggon was common, but coaches and waggons were now all driven off the road.

Edward Bury, of the Liverpool firm of Bury, Curtis and Kenedy, was the company's locomotive engineer. We very seldom saw him, and he did not very often look round the shops. He took care that Bury, Curtis and Kenedy should make a good thing out of the company. The locomotive plant consisted of fifty-five engines, nearly all made by Bury, Curtis and Kenedy at the Clarence Foundry in Liverpool. There was an iron foundry in the Wolverton works, but it was never used, for Bury, Curtis and Kenedy furnished all the castings. Wolverton was simply a repairing shop, and every cylinder, every eccentric, everything in cast iron, was supplied by Bury, Curtis and Kenedy. No railway company at that time built its own engines. The locomotives on the London and Birmingham were small and light compared with those now in use. A few used in the goods department were coupled. The passenger-trains were run by an engine on four wheels, the driving-wheel being about five feet six in diameter. They were swift, but hardly strong enough for the work, and many of the trains required two engines to draw them, and a pilot engine was always on the station at Wolverton ready to go in search of belated trains, and assist them. The Gifford feeder being unknown, the only way of supplying the boiler was the pump; and many was the time the pilot engine, as the steam blew off and the water got less, was trotted up and down a mile or two that its pumps might supply it. Some time afterwards came the device of two blind wheels stationed under the line, their unflanged ones forming a few inches of the line itself.

Thus the engine was slowly drawn until its driving wheels rested on the blind wheels, and then it could pump away without moving; but I believe that to get it off or on the blind wheels it had to be pushed with a crossbar.

It was not the infancy, but still the childhood, of locomotion, and Bury, Curtis and Kened having their drawings, templates, and patterns, were in no hurry to introduce improvements.

We were in all about five hundred hands, a mixed group of Londoners, Lancashiremen, Yorkshiremen Scotchmen, Welshmen, Irishmen; and among us was a man of great stature and magnificent proportions, Polish gentleman, who kept himself very much to himself, and was very taciturn. Our foreman was a Scotchman, named Patch, who afterwards went to be superintendent on the Edinburgh and Glasgow line. I knew most of~ the men, but of the five hundred I don't think I knew~ more than a dozen who went to church or chapel. Those who in these times so speak of the working men as to produce the impression that they have fallen away from religious ordinances are very much mistaken. Of these in Wolverton, with abundance of church accommodation not far off~ with no counter attraction but the fields, with no reason to complain of being shut up all the week in the close and unhealthy atmosphere of' smoke, not more than two per cent. ever went to worship. Are things worse now? I very much doubt it, There was, however, very little to induce us to go to either church or chapel. Most of the neighbouring clergy were gentlemen who followed the hounds. The parson at Stony Stratford had spent some years in prison as an insolvent debtor; the remembrance of which must have been strong upon him, for as he droned through the service, he grew animated and earnest in praying that those evils which the craft and subtlety of the devil or *man* worketh against us be brought to nought. All that I can remember of the old chap is his extreme stupidity and dulness, and that loud emphasis upon the word 'man.' Many of the clergy were very hopelessly in debt, and were held in very little esteem. There was not one for ten miles round who could preach so as to interest any mortal creature. One of them, a great fox~hunter, was a magistrate, who occasionally fined a station-man for trespass or for poaching, and, of course, was hated and cursed by all the station-men. The attendance at the churches was wretched; the station-men were not the only men who did not go to church. Not one farmer or farm-servant out of ten was often to be seen within the consecrated walls. The congregation at Old Wolverton Church was seldom a score, and considering what a dismal fool the parson was, I wonder there were so many. He seldom preached; there was some attempt at reading a sermon once in two or three weeks, and it was once too often; and he mumbled the incomparable Liturgy in a most atrocious manner, the object evidently being to get through the thing as soon as he could. There was only one service on the Sunday, and in this good old style, as in the Isle of Man, the parson and the clerk had it all to themselves.

The state of Nonconformity in the district was not much better. At Newport Pagnell there was an Independent Minister, Mr. Bull; but Newport is four miles from Wolverton. At Stony Stratford there were two small 'interests,' an Independent and a Baptist. The Independent Minister sang mournfully through his nose, and was very dull and prosy; the Baptist Minister, Mr. Forster, was a man of considerable abilities and a good preacher. There was a Methodist Chapel also in the town; but it was only a poor little thatched cottage. There was nothing attractive about any of these places.

It was in the fine summer weather of August 1840 that I went to Wolverton, having a second-class pass from Liverpool. I arrived on Friday evening, and went into the erecting-shed on Saturday morning; and at the dinner-hour had to pay my footing. This was done at the vile public-house close to the station—a house which went by the name of ' Hell's Kitchen,' a name it well deserved. An old proverb says, that if an Englishman settled on an uninhabited island the first building he would put up would be a public-house. A public-house was the first thing built at Wolverton by the directors of the London and Birmingham There was no church, no school, no reading-room ; but there was ' Hell's Kitchen.' And in that 'Hell's Kitchen,' that first afternoon, I had to pay about ten shillings for drink. 'Hell's Kitchen' was a horrid place; always full of mechanics, navvies, labourers, tramps of all kinds; at least one hundred station-men spent there half the dinner-hour and perhaps half their wages. Working men drank just as hard in those days as they do now. That afternoon, as I came up from 'Hell's Kitchen,' I was very much disgusted. I had taken no drink myself. I saw a workman leaning against a paling, who said to me in most unmistakable Cockney, "Well, mate, what are you going to do with yourself to-morrow?" I answered that I did not know, and did not care. "Well," he said, "I am going over to the Independent Chapel at Stratford; will you go along with me?" I said I would, and we went. That good man was John Page, an engine-fitter, a godly character, a man who said very little about religion; but that quiet invitation to go with him˜ unaccompanied by any cant had a quiet power. It had much to do with settling me into good habits ; it made me the companion of a good man, and saved me from other and very different company. I used to lodge with Page at the house of an old man named Stoney in Old Wolverton. Old Stoney had a son named Edmund, who worked in some capacity at the station. I lodged in his father's cottage for but two or three months, and then lost sight of him. But this year (1879) I had a letter from him. I had, of course, all but forgotten him, having heard nothing of him for thirty-nine years. He informed me that he was a grocer in Sheffield, whence his letter was dated ; reminded me of our old acquaintance, spoke of his business as good and prospering, and would be delighted to see me and have me as his guest when next I might be in Sheffield. I replied, thanking him. In less than a month he wrote again, asking me to lend him fifty pounds to save him from having the bailiffs put into his house. Now I saw why he had written to renew the acquaintance!

My wages at Wolverton were for the first year 4s. a week, 5s. for the second, 6s. for the third; for the remainder of my needs I had to draw upon my father's scanty means. At that time he had nine children to support upon less than £200 a year; my brother Robert in Liverpool being still the only one in part able to keep himself. The tools employed in the works were of a very simple character in comparison with those now in use. There was no steam-hammer; all the forging was done by hand, and it was a fine sight to see seven or eight stalwart strikers, at the forging of a crank-axle, plant their huge hammers in rapid succession upon the spot indicated by the smith with a piece of rivet rod-iron. We had no travelling-drill; all key-ways had to be first drilled in round holes, and then cut with a crosscut chisel, and finished with the file. With the exception of a small machine for cutting nuts, we had no shaping machine of any account. Every surface that could not be formed by the planing-machine had to be chipped and filed. All the light turning was done by hand, without a slide-rest. Altogether, the work of an engine-shop was much more laborious than it is now, and required much more skill. Machinery has to a large extent superseded both muscle and brain, and a boy set to a machine can do more and better work than would be done by a skilful mechanic. Yet there were men who could do wonderfully true work. I have seen a fitter take two rough pieces of wrought iron of more than one pound weight each. I have seen him chip them to a surface almost perfectly smooth, and then with files so perfect the surface that when placed one upon the other the lower piece would hang to the upper by the force of molecular attraction, as if glued to it. Of course I do not mean that they were so fast joined as glued surfaces, but it required a sensible effort to separate them. I never could do anything like that; in fact, I was but an indifferent workman. My best performances were at the hand lathe ; there I did pretty well. Our best mechanics were, I think, the London men; one of whom, named Airey, a relation of the Astronomer-Royal, could turn out an extraordinary amount of good work. The Lancashire men came next, and I think the Scotchmen were the worst. The fitters' wages varied from 28s. to 33s. per week; smiths about 30s.; labourers, 16s. to 18s.

Chapter X My Mates at Wolverton.

I CANNOT say that we worked very hard. It was an idle shop. In the fitting-shop we took our turn to watch for the foreman as he came up the sheds, and the word 'nix' saw every man and boy at his place. The lathes and planing-machines had been moving, hut doing nothing else; now the tools were thrown into gear, and all was work and bustle. But we had some shameful idling. I am sure that every set of taps and dies made by Figg the tool-maker must have cost more than its weight in silver. And Mr. Figg was a local methodist preacher, who always wore a white shirt and black trousers, and was quite a swell. Another man who made cross-heads rarely did an hour's work in a day. Most of his time was spent in arguing in favour of socialism, and turning the Bible into ridicule. There were a considerable number of knobsticks in the shops; men who had gone in on strikes in London and elsewhere. "What are you talking to that fellow for, don't you know that he's a — knobstick ?" So said a society man to me, when he saw me conversing with a shopmate whose knobstickism I was not aware of. There was much angry disputing, much sullenness, much hatred, hard words, and occasionally hard blows over this knobsticlc business. Yet the knobsticks were as a rule good workmen. Airey was one, and so was Cole (called from his cadaverous look, Captain Death), the only man to whom was entrusted the difficult job of -turning the crank-axles. Among our men was one from Birmingham, a brother of a well-known atheist. He was a particularly poor creature in mind and body, and held his brother's opinions. We had a number of infidels among us. One of these died avowing his atheism to the last in the most horribly profane manner. A few of us went before the funeral to ask the parson old Quarlby, whether he could omit the expression "in sure and certain hope" as he read the funeral service. We told him that the whole band of infidels would be there to sneer and to triumph, as poor Alec was honoured with Christian burial. Old Quartley had just come in from the hunt, and was taking his boots off. His reply was," Lord love ye, but what the devil would the bishop say to me if I did as you wish?" So the atheist was buried in sure and certain hope of a glorious resurrection, and three of our Churchmen from that hour were Non conformists. The talk in the shops was for the most part profane and beastly in the extreme; the least profane and beastly among us, with the exception of the few Christians, were some of the infidels, one of whom, a Scotch-man named William Angus, was one of the finest men whom I ever knew. The Shorter Catechism had made him an unbeliever. The leading infidels at Wolverton and the chief drunkards were Scotchmen. I remember very little conversation upon industrial subjects, and although it was a time of great political agitation in regard to Chartism and Free Trade, there did not seem to be a spark of political intelligence or spirit among us. We never saw a newspaper, excepting the 'Weekly Despatch,' a few copies of which came on Sunday morning, but that was read chiefly for the prize-fights and other sporting intelligence which it contained. I should think that one-third of the men were unable to read a single word, and I often wrote their letters for them, and read for them letters they received.

Four of us, Edward Hayes, William Harvey, William Mickle, and myself, drew together. They were journey men, but young, the oldest not more than twenty-five years. We agreed to lodge together with a peasant named Cox at Old Wolverton. Hayes was a little man, a clever, skilful workman ; he came from Manchester, and was great in phrenology, and in Combe's 'Constitution of Man.' Harvey was a Derbyshire man, one of the best workmen in the place, and gifted with a dry and pleasant humour. Mickle was a Scotchman, brought up in London; a boisterous but kindly fellow, whom Hayes pronounced to be a man in whom combativeness and self-esteem were abnormally developed. The four of us slept in two beds placed in one small room. We had our meals in the lower room of the cottage, which was the kitchen, and there was a small room, about eight feet square, which we converted into a study, and in which we tried in the evenings to improve our minds, which, sooth to say, sorely needed improvement. On Sundays, Hayes generally went out into the fields to meditate; Harvey went to the Methodist Chapel at Stratford; Mickle wandered from one place of worship to another; and I went to church somewhere in the neighbourhood, generally to Stratford, because there was an organ there, which, however, was very execrably played. Our studies were various. Hayes went in for philosophy; Harvey for theology; Mickle for mechanics ; I for mathematics. I don't think we read a novel all the time we were together, and our whole stock of books was not worth £5.

The three years passed with very few incidents to break the monotony of our lives. We rose at half past five and walked to the works a mile off, cooked our breakfasts at one of the forges in the smiths' shop went home to dinner at one, returned at two, and the bell rang again at half-past five; on Saturdays at four; in all 58½ hours per week, with every evening free. I think far too easy work.

One event of these years I cannot forget. It was my first visit to London at Christmas 1841. I well remember the lurid glare in the sky as in the winter evening we approached the great metropolis. I had only three days in London, but I used them in seeing all that I could see. Among other places I got into the ball of St. Paul's, where I found three sailors who insisted on my drinking some of their rum, the effect of which threatened to make my going down much more rapid than my going up. No place known to me has altered more than London in my time. The alterations have not all been improvements, i. e the railway-bridges over the Thames.

Among my excursions, for which there was no time but Sunday, was one to Olney. In the neighbourhood there is or was a fine old tree called 'Cowper's Oak,' hollow, with a seat in it. I met an old man who well remembered the poet sitting there. I once went with Mickle to Northampton, walking thither on Saturday evening, and walking back on Sunday evening. In the afternoon we heard a Mormon preacher haranguing in the Market Place. Mickle, the combative, attacked him, and had a long wrangle with him, in which I joined. It was my first attempt at public speaking. Strange to say, thirty years afterwards, I heard the same

103

Mormon deliver the same sermon (or one very much like it, and on the same text) in the Tabernacle in Salt Lake City.

Chapter XI I Become a Teetotaller and Sunday School Teacher

About this time I became a teetotaler and a Rechabite, as did my fellow-lodgers. This brought me into great disfavour with the drinking workmen, and I was commonly called "a — teetotaler." The teetotalism led to my going to temperance meetings. The first in which I took part was held at the village of Daneshanger, where in attempting to make a speech I utterly broke down in confusion. Temperance meetings were scenes of great interruption and uproar: for attempting to persuade people to sobriety we were persecuted, hooted out of the villages, and pelted with mud. At the station close to the canal bank there was a small temperance coffee-house, kept by a man named Spinks. A few of us thought that we might hold a Sunday School there. We obtained the use of the room, and started the Sunday School, and I taught there on Sunday forenoon and afternoon for some time. That was the first and for more than a year the only religious service of any kind in Wolverton. Neither clergy nor Nonconformist ministers took the slightest interest in the people, altogether numbering about one thousand, that were gathered about the station. That was my first Sunday School teaching, and my last. I disliked the work immensely; I had no gift for it. It is my duty to take an interest in Sunday Schools, and I have had for a great many years as good Sunday Schools as any in Liverpool; but I have always been thankful that no more than an occasional looking in has been required of me, for a Sunday School, and indeed any school, is no pleasure to me to this day.

It is proverbial that corporations have no conscience. For a time this seemed true of the London and Birmingham Railway. They had brought a large number of men, women, and children to Wolverton, and children were increasing in number rapidly. There was no school of any sort within two miles. The nearest were the National Schools at Stony Stratford, kept by the clerk of the church, a very drunken rascal, and that school was hardly large enough for the requirements of the town. But for years the Company made no school provision, and the children at the station were growing up in utter ignorance. At last the Company proved to have somewhat of a conscience, and they built a British School, a poor shabby

104

thing it was, in keeping with the abominable cottages, which they built for the workmen. The Company's conscience was indeed so exercised that they sent a parson, a Church of England man, to hold service and preach in the school. This was the Rev. George Weight; he had been one of Rowland Hill's people; had, I believe, been assistant to him in his last days at Surrey Chapel, but became a Churchman, and cordially hated Dissent. But he was a thorough evangelical, and a capital preacher. I went to the school on the first Sunday of Mr. Weight's services, and a good many of the men attended, moved by curiosity. There was at Wolverton a drunken Scotchman named Dan Rintoul; he was considered the most intemperate man on the place. He was at the service, and as usual was very drunk.

He had probably never before seen a clergyman dressed in a surplice. When Mr. Weight, who was very precise in his ecclesiastical vestments, made his appearance, Dan cried out, "You fool, go and put your breeches on, and don't come here in your shirt." Dan was soon hustled out of the place by Bill Webber, a little Cockney knobstick, who acted as clerk. That was the beginning of public worship in Wolverton. Dan was taken before the magistrates and fined heavily and of course neither he nor any of his pals ever went to church again. The Company soon after built a small church at the station, but I don't think that more than a dozen station-men ever entered the place. It was a great mistake to appoint a knobstick as clerk, that of itself was quite enough to keep three-fourths of the men away.

I introduced myself to Mr. Weight as the son of a clergyman, and found him very friendly. I had already begun to cherish some thought of becoming a parson, and Mr. Weight encouraged me in the project, and proposed to give me lessons in Greek, of which I was wholly ignorant. I procured a Greek Grammar, Lexicon, and Testament, and most of my evenings were spent in this new study. As 1 have said, Wolverton was an idle workshop. At that time I was engaged for some weeks or months in tubing boilers, and I generally took the fire-box end, having a very lazy mate at the smoke-box end on the look-out for the foreman, whose approach was signalled to me with a stroke of the hammer on the boiler. I did a good deal of study, and by the light of a candle wrote my earliest Greek exercises on the sides of the fire-box with a piece of chalk.

Chapter XII I Decide to Become a Minister of Religion

TOWARDS the end of 1843 I ventured to write to my father, telling him of my wish to enter the ministry. He very reluctantly gave his consent. He never was hearty about it, and would much have preferred my continuing at my trade. I don't wonder at it. The ministry had been to him a life of poverty and hard ship, and, as I have already hinted, he was only half a Churchman. My mother, however, viewed the case more favourably. She had always wished me to be a minister, and my going to business was a sore disap pointment to her. And so it was arranged that I should leave Wolverton at the end of the year, and return to the Isle of Man, and go to King William's College.

The end of the year came, and I left Wolverton and went home. I have no doubt that the three and a half years' intercourse with so many working men from all parts of the country has proved of great advantage to me in my ministry. The practice of speaking acquired at the temperance meetings was also a great help, and my practice of going as much to chapel as to church gave my religious views and sympathies a breadth which is perhaps unusual at so early a period of life. Yet I had often thought that it was cruel towards my parents to take the step I took. My restlessness had given them much trouble already; I was in my twenty-first year, and was still a burden to them. Receiving 7s. a week and sometimes 10s. (when we worked over time), that burden was being lightened. In another year, perhaps half a year, I should have been earning my own living, to their great relief; and yet in this new notion of becoming a minister I threw myself again upon them. I think my father had good reason to grumble, and I have sometimes wished that he had finally refused his consent, and kept me to my work, for I was not so vain as to suppose that God's cause could not have gone on quite as well had I never entered the ministry. Here the proper and orthodox way would be to cant about an inward call to the ministry, a burning love of souls, an intense desire to consecrate myself to the service of Christ. Of such things I say nothing. I never felt that the Lord had need of me, as he once had of one disciple at Jerusalem. I suppose I called myself to the ministry urged by various motives, and if all men who speak upon this subject spoke in honesty and good sense, I think they would say much the same thing as I say here. The stories of men whom I have heard at ordinances and settlements, about the Lord having called them and led them, have often turned out fictions, or something worse. Before bidding farewell to Wolverton, I may add that, while the Company did provide a school and a church, they did not, until after I had left, provide anything in the nature of a Mechanics' Institution or Reading-room. They did not want the men to improve their minds. Not a single thing did the Company for the amusements of their men save and except 'Hell's Kitchen.' They did a year or two afterwards build a Mechanics' Institute, but it was a poor, mean, shabby concern ; an utter disgrace to them.

I left Wolverton for Douglas in the beginning of 1844, to study for the ministry of the Established Church. On reaching home, I found my father in no way gratified by the resolution I had taken, and, as stated above, I do not wonder at his dissatisfaction and want of cordiality, for I felt that I ought not to have cast myself upon him for support. It was intended that I should go to King William's College, where, as the son of a Manx clergyman, I would have instruction gratis. However, the greater part of the College was burnt down, and the arrangements of the Institution so disturbed, that I could not be conveniently received, and it was thought best for me to study with my father until August, when the College would begin its second half-year. From January until August there was I at home; but never did any poor wretch feel himself less at home. Indeed I think that home ceases to be home when a fellow has reached his twentieth year; he had far better be somewhere else; he is a man without a man's freedom. I felt this horribly. My father was morose, feeling, no doubt, that he had done enough for me, and that I ought to be on my own hook. I read with him, chiefly Latin; thrashing away at the Greek by myself; for of Greek I suspect my father was innocent, as he had never looked at it since he left Castletown Academy. I have no wish to dwell upon those miserable seven months.

The seven months previous to my going to the College were spent in very diligent study, and in August I went to King William's, lodging in the house of Mrs. Kewley, on the Green, Castletown. The Principal of the College was the Rev. Robert Dixon, known by the lads as 'Bobby,' a Cambridge man. My chief business with him was in the Greek Testament and other Greek—Thucydicles and Homer, Sophocles, Euripides. The Greek Testament used was Bloomfield's. The second master was the Rev. J. G. Cumming, who had scientific tastes, and was considered an authority upon the geology of the Isle of Man. Under him I read Latin, chiefly Sallust, Virgil, and Horace. I did something, but not much, in mathematics. I ought to say that, under Dixon, most of us intended for the Church dabbled a little in Hebrew. I certainly worked hard all the time I was at King William's; rising at six in the morning, reading a great deal, and preparing for the classes and lectures. But it was not a school for me. I wish I could have gone to Oxford or Cambridge, but that was hopeless. I made the best of such chances as I had.

The society in and around Castletown was generally good. We were fairly well fed and cared for by Mrs. Kewley, there being about a dozen College lads in the house. The clergyman of Castletown was Mr. Parsons, a kindly, hospitable old fellow, and I and some others of the College were often asked to his house. There was a company of soldiers in Castletown, the garrison for the whole island. The soldiers went to Parsons' church. The Government chaplain, John Howard, was at one time his curate, and preaching rather longer than usual one Sunday morning, Parsons, looking up from the desk to the pulpit, said, "John, John, come to a conclusion the Governor's dinner is getting cold." The beadle's name was Buchan, and popular rumour fastened upon Parsons the old Joe Miller, that a dog came into the church and yelped while he was reading prayers, "That it may please thee to bless Adelaide the Queen Dowager, the Prince Albert, Buchan put that dog out, and all the Royal Family."

While I was at Castletown there came a new governor, Governor Hope. I was present at his installation in the Castle. In the name of ' Victoria Rex,' silence was commanded the Governor took his oath that he would administer the law as evenly as the back bone doth lie in the herring.

Building a Community

London and Birmingham Railway Board
Euston: 11th June 1841

Read report of Sub-Committee for the management of the Wolverton Schools.

1625: Resolved that the appointment of Mr. laing on trial as Master of the Wolverton School, at a Salary of £100 per annum, together with a House and firing, be confirmed.

1626: Resolved that the Chairman be requested to express to the Radcliffe Trustees the wish of the Directors that no time should e lost in the appointment of a Minister for Wolverton:- and that, until the Church is erected the Committee be authorized to appropriate a room in the School for the performance of Divine Service.

Euston: 9th. July 1841

The Chairman reported that communications with the Radcliffe Trustees with reference to the appointment of a Minister for the Church at Wolverton, and that Mr. Bramwell required on behalf of the Trustees that the Clergyman should have sole control of the School, the Company to contribute £50 per annum towards the stipend.

1659: Resolved that the sum of £50 be contributed annually towards the stipend of the Clergyman to be appointed by the Radcliffe Trustees for the Church Service at Wolverton.

Read letter from Mr. Pousett, 8th. inst.

1682: Resolved that the sum of £25 be expended in the purchase of Books for the Reading Room at Wolverton under the direction of the Committee for Superintending the Schools, and that Mr. Pousett be desired to send for their guidance a Catalogue of the Books now belonging to the Reading Room.

The Radcliffe Trust

Minute of the meeting of 11th Jun 1842

The Trustees again directed their attention to the peculiarly distressing state of the large assemblage of persons who are attracted to the Wolverton Station by the extensive commercial operations of the London & Birmingham Railway Company but are unhappily destitute of the means of receiving adequate spiritual instruction in consequence of there not having been as yet provided any sufficient place of worship.

This circumstance having led the Trustees to revert to the subject regarding the erection of a church or episcopal chapel and a minister's residence, on a site contiguous to the railway, they feel it a duty incumbent upon them to make a renewed representation to the Directors of the Railway Company and to refer to the resolutions of the Trustees dated 8th June 1840, a copy of which were at that time transmitted to the Directors, by which the Trustees declared their willingness to provide a site for a chapel, for a Minister's residence and for a burying ground, as well as a permanent endowment for a Minister, and moreover to defray hereafter the expense of repairing the chapel and Minister's House.

To this offer the Trustees added the expression of their hope that the costs of erecting the Chapel and a Minister's House would be provided for by the London & Birmingham Railway Company out of their funds or by voluntary contributions.

The Trustees observe with regret that little has yet been done to meet the wants of the 1,500 persons at present representing the population of the Station at Wolverton.

It appears that since the meeting of the Trustees in June 1841 and in consequence of the Resolutions then entered into, the Revd. George Weight has been nominated and licensed by the Bishop of Lincoln as the Chaplain of the Station.

That a School Room capable of holding 250 persons has been there fitted up as a temporary place for the performance of divine service, but it is found to be very inconvenient and quite inadequate for the purpose.

It is manifest therefore to the Trustees that every effort ought to be made to remove the evil by providing a becoming and suitable place of worship to be effected by building a plain but substantial chapel capable of holding 600 persons with a burial ground attached thereto and a house for the residence of the Minister.

The Trustees cannot but entertain the belief as well as hope that the Railway Company will participate in this sentiment and will feel that independently of religious considerations, it would be even in a merely secular point of view most advantageous that the population which have settled at the Station should have afforded to them the comforts of religious consolation and the benefit of receiving such spiritual instruction as is deemed to be essential even in the smallest and least populous parishes.

The Trustees have therefore determined to make a proposal to the Directors of the Railway, the acceptance of which will enable them with greater confidence to apply to the Court of Chancery for permission to devote a proportion of their Trust Funds to the accomplishment of so great and necessary an object.

The Trustees calculate that the sum of £4,000 will be sufficient to build the Chapel, the Minister's House, and the wall surrounding the burying ground.

In addition therefore to what the Trustees expressed their willingness to do towards the attainment of these purposes .... they now propose to appropriate £2,000 out of the Trust Funds towards a Building Fund, and earnestly invite the Railway Company out of their Corporate Funds or by private subscriptions to contribute a similar sum with the assurance that as soon as the Railway Company are prepared to lodge in the hands of a Banker £2,000 the Trustees will immediately make an application to the Court of Chancery to sanction the important object in contemplation.

Consecration of Wolverton Church

The Times Wednesday May 29th 1844

The courtesy of the directors of the London and Birmingham Railway Company enabled us yesterday to be present at the consecration of the church which has just been erected at their station at Wolverton. Connected with the performance of this ceremony were some circumstances of more than ordinary interest, and we may therefore be excused for dwelling on it in some detail.

Many travellers upon the Birmingham Railway are not aware that there is anything more remarkable at Wolverton than its commodious and well-supplied refreshment room. This error is perfectly excusable, for until within a few years Wolverton was nothing more than a farm, the property, we believe, of the Radcliffe Trustees. A consequence of the railway, however, is the settlement of a colony upon this somewhat remote spot – a colony of engineers and mechanics all constantly and regularly employed by the Leviathan Company which gave birth to the town, and which has invested it with sufficient importance to entitle it a place upon the map of England. The circumstance under which the town was called into existence may be worth relating. When the Birmingham Company's bill was first introduced to the notice of Parliament it was proposed to establish a central station at Northampton, a town which, from its own importance and its central position upon the contemplated line, appeared to be a most eligible position for the Company's works. The shortsightedness of the Northampton people, all at that time engaged or interested in coach traffic, prevented the perfecting of the arrangements. After a vast deal of opposition, attended with great expense to all parties, they succeeded in forcing the Company to abandon their project, and select another spot on which to carry on their works. As there was no other town of sufficient importance eligibly situate on the route, the managers wisely sought a counterbalance for the disadvantage. They saw that if they lost some facilities by placing their station remote from a town, they would gain by the increased steadiness and regularity of their workpeople. Accordingly, Wolverton, a healthy spot, many miles from any place of public resort, was selected as a site for a large station, and there, as we said before, the Company have founded a colony of engineers, which is rapidly flourishing while Northampton is going to decay.

At the present time Wolverton is a neat, brick-built, clean little town of eight or ten streets, regularly and well laid out, containing houses of different classes,the smallest of which, however, are superior to the general run of mechanic's dwellings in our large manufacturing towns. We saw Wolverton, no doubt under great advantage yesterday but there are at all times some unfailing tests of the character of a population, and the general cleanliness of the interiors, and the absence of beer shops, may, in this case, be regarded as two most favourable symptoms. The present population of the place is, we believe, 1,500. Of this number there are only a few families who are not employed by the Company. The houses, which have all been built by the Company, are let to their servants at a very moderate rental; and as a reward fot that description of labour in which the inhabitants are principally engaged is highly remunerative, it may safely be said that absolute poverty is unknown in Wolverton.

The works at Wolverton, which are so placed upon the line as almost entirely to conceal the town, are, as may be supposed, very extensive. It would occupy too much space to dilate upon this topic,but it may be incidentally mentioned that one of the engine houses is capable of containing 24 to 30 engines; that in the factory there is always one new locomotive in progress, whilst the number under repair is of course considerable. In this little place alone, we should assume that this company must spend in wages, &c., from £40,000 to £50,000 per annum.

Some time since the Company, regardful of the spiritual as well as of the temporal condition of those who are employed in carrying on the schemes of profit, erected a school in the town, which is conducted on excellent principles, and at which about 100 boys and girls are in daily attendance. No less mindful of the soul's welfare of the elder than of the junior branches of their little community, the Company have now erected a church, with a parsonage house attached, upon a site liberally given them for the purpose by the trustees of the Radcliffe estate. The church itself is a plain building, in the Saxon style of architecture, which has lately become popular. Perhaps it is not a particularly good specimen of its class, but its want of exterior attraction is compensated by its commodious interior fittings, which provide accommodation for nearly 1,000 persons. All the seats, we rejoice to say, are free: there is not a single pew or reserved seat in the fabric.

The ceremony consecrating the building was performed by the Lord Bishop of Lincoln, the diocesan, in whom, we believe, the right of presentation to the church is to be vested. His Lordship was accompanied to the church by a large body of his clergy, and a number of gentlemen of the vicinity, and others interested in the work. Mr. T. B. Escourt, M.P.; Mr. George Carr Glyn, the Chairman of the London and Birmingham Company; and several of the most influential directors. The church was crowded in every part, and the entire congregation appeared to be most respectable. The Bishop preached an eloquent discourse from the 10th chapter of Isaiah and 47th verse, and took occasion in the course of his sermon to dwell upon the circumstances under which the church was erected; and to exhort those amongst his hearers who were engaged in solving the profundities of physical science not to neglect the study of the natural sciences which led to reflection upon the greatness of God and the comparative littleness of man.

After the ceremony a numerous company partook of a very elegant *déjeûner*, served in the large room of the station, the Chairman of the Board of Directors presiding.

A Letter from George Bramwell to the The Times

To the Editor of The Times Saturday, Jun 1st 1844

Sir, On reading in your valuable journal of the 29th inst. An account of the consecration of the above church, I perceive there are some errors into which you have inadvertently fallen.

It is stated that the Radcliffe trustees, the owners of the Wolverton estate, liberally gave the ground for the new church and parsonage, but, in truth, they have done a great deal more, or the church would never have been erected.

The Radcliffe trustees have paid the entire expenses of the erection of the new church and parsonage, which will amount to about £5,000; beyond this, they pay £100 a year to the minister of the church towards his stipend.

The Birmingham Railway Company have, doubtless, given some assistance towards this good work, having, by subscription, raised £2,000, which has been appropriated for the endowment of the minister.

I trust you will allow this explanation a place in your columns.

I remain, Sir,
Your most obedient servant.
GEORGE BRAMWELL
Solicitor to the Radcliffe Trustees.

Furnival's Inn, May 30th, 1844

P.S. The patronage of the church is vested in the Radcliffe trustees.

Letter from George Bramwell to Ralph Cartwright
Furnivals Inn
Nov 16th 1843

Dear Sir,
I was at Wolverton last Tuesday and the Mr. Weight complained to me of the disgraceful scenes at the "Radcliffe Arms" Public House – I have (on

receipt of your letter) written a strong letter to Congreve and Clare[4] on the Subject – to remove the present under tenant and to put in a man who will conduct the house in an orderly manner.

The application of Mr. Glyn[5] for more land surprises me for there is a considerable portion of the land sold to the Company unbuilt on and now used for garden ground and room enough for many houses and streets.

You are quite right in throwing such a doubt as to the power of the Trustees to grant building Leases. I do not think the Trustees wd grant a valid building lease without the sanction of Chancery or perhaps a private Act of Parliament wd be found necessary.

The Church at Wolv. Is in a very forward state: the exterior is quite finished and altho' the Spire has excited criticism (both as to its small dimensions and position) still the Church is admirably built & the style of the work reflects great credit on Grissell & Peto.[6]

Letter from the Rev George Weight to Geo. Bramwell

Wolverton

June 15th 1846

Dear Sir,

You enquire abt. The state of the Congregation here, especially in the morning.

It is not good. You ask the reasons –

The general reason wd. be as truly applicable to every rural Church in England, as to this. Namely, that the men, under the plea of having worked hard all week, indulge in sloth and idleness on the Sunday Mgs. The women keep at home to cook the Dinners.

Another reason here is, that there are so many village feasts on the Sunday, within reach of our people.

Another reason is the Irish Roman Catholics, & various sects of Dissenters. But I believe the true and full reason is,that our people are so unsettled. For example; last year we suppose that, at least 300 men with their wives & families left us; &, of course, 300 stranger came in their place. Of the latter 300 I suppose not 50 are here now. To be told that 20

---

4 John Congreve and Joseph Clare, lessees of the Radcliffe Arms.

5 Chairman of the London and Birmingham Railway Company.

6 Architects.

families had left on any day, wd. give me no surprise. In such a state of things it wd. be manifestly impossible for any Clergyman to do much for his flock.

Wed J. 17th.

I was called from home yesterday, on affairs connected with the schools, or this wd. have been sent off.

Three years ago we had an excellent congregation, & a noble staff of Sunday Sch. Teachers. Then came the Change among the people, & I have had much of my former work to do over & over again. Now, before I can get the people into good habits they leave us. I lost three of my Sunday School Teachers, last Monday –

In order to give you an exact idea of the State of the Congregation I desired four persons last Sunday to count the people as they left the Church. There were,

20 years old & over	73
Under 20	30
School Children	130
In the morning	233

In the evening of all ages – 201. Only a few of the children, perhaps 50.

No village for scores of miles round us wd. give a better result, considering the singular character of our people –

You enquire also respecting the Reading Room –

They assure me that they do not open it till one on Sunday, and close it at six.

The most objectionable paper they now take, so far as I know, is the (Sunday) Weekly Dispatch. This vile paper is much read here. They used to take The Satirist also.

Everything in the Reading Room is managed by a committee; every member votes, & the majority decided. Some of the religious people have left the room in disgust, so that the Sunday party manage as they please.

My exertions for the good of the Company's servants are by no means confined to this place. I have about seven hundred volumes of books in circulation between London & Birm., & these are all lent free, to any servant who chooses to apply for them.

I enclose a copy of an inscription on a silver Tea Service lately presented to me by my flock here.

We have let 74 seats in the Church, at 1/- each per quarter. We have 30 Communicants, & the Sacrament every month.

When we get a more steady & settled people we shall I hope, with God's blessing, do more good.

I am, Dear Sir,
Very truly Yours,
Geo. Weight

The Expansion of Wolverton

The new town grew rapidly. Within a few years the population has gone from none to well over 1.000. The operational demand of the railway now required more workshops. The Trustees, by now not on good terms with the railway company, were not sympathetic to an increase in the size of Wolverton, by now overtaking Stony Stratford in population, posed serious threats to the customary rural way of life. A man called Edwin Driver was commissioned to make the following report, written in longhand and now in the Bodleian Library, Oxford.

Report of the Application made by the North Western Railway Company for permission to purchase a further quantity of land in the parish of Wolverton for the increase of their Works near the present station there, from the Estate belonging to the Trustees of the Radcliffe Charity.

Upon taking into my consideration this Application, I have entered into the subject with the proper spirit which I have reason to believe is the wish of the Trustees for this Charity should govern their decision, namely, that of not withholding their means of rendering every reasonable & proper assistance to this large and important Company to facilitate their improvements, but with a perfect recollection of some impressions which were left upon my mind upon my recent survey in 1844 of this fine property at Wolverton belonging to the Radcliffe Charity touching the inconvenience, annoyance and prejudice, which the Company establishment at Wolverton had unavoidably occasioned.

I feel it my duty to revert to these here and bring them forward for the particular consideration of the trustees, & to suggest such measures as a remedy, so far as they may apply under the circumstance as appears in my judgement what the Trustees will be. fully justified in calling upon the North western railway Company to comply with.

The present application, according to the plan furnished by them, states the further quantity now desired, to be 10 1 6, but to carry out which, deeming t necessary or the due protection of the remainder of the Charity Estate, and having, upon this occasion, been to Wolverton to take another inspection of the property, I am of the opinion the Trustees should compel the Company to make some further purchases of some adjoining lands and I annex to this report a copy from the plan showing the whole that I propose to refer to distinguishing the several parcels by colours, for considering the very large increased population, which has necessarily resulted from this large establishment renders it absolutely requisite that the ambit thereof should not only be most accurately, but well defined, but also confined as fa as practicable by substantial Walls or other satisfactory Bound fences, so as to prevent , as far as possible, any occurring or recurring injury or annoyance from trespass or further encroachment from the Inhabitants or others.

The Grand Junction Canal, which passes at present round two sides, forms a satisfactory boundary so far as it extends and abuts thereto, but on the other sides or portions, the external Boundaries, at least where particularly marked on the accompanying plan, & shown by a double red line, nothing short of such a Boundary being defined by a strong built stone or brick wall, 10 ft high above the surface of the ground, can effectively prevent a constant annoyance by trespass into the adjoining Fields and Premises. . No stronger proof or evidence of the objectionable intrusion or Encroachment which may be apprehended & for want of such defined external boundary fence need be advanced, than the existing trespass which has allowed to be carried on from the building on the west side of the Company's present works. The boundary at this part was a straight line running from the public road near the Royal Engineer public house in a north direction down to the Grand Junction canal by the side of the Fences where is a row of trees on the west side of the boundary.

A piece of ground Nos 179 & 180 containing together 8 3 26 was afterwards applied for being let out as Cottager's Gardens to the Railroad Company, which I do not object to, as this is a very laudable object, but it was then found desirable I suppose, to have a load let out on the east side for convenience of the occupiers. The consequence has been that a long street of dwelling houses have been built by the Company against the western boundary of these premises, each having a small outlet in the yard and washhouse also attached, and so soon as such adjoining road was made for the accommodation of the occupiers of these gardens, a back gate has been opened from the rear of all those houses to lead into such road which is a most objectionable encroachment & nuisance I ought not to have ben permitted, but appears to have been quietly submitted to, as I have not been informed of any objections having been raised or provisions made for payment of any satisfaction which is a precaution which certainly should have been attended to, and in neglect thereof the system of encroachment has been extensively carried on by most or at least very many of the occupiers over this private road by erecting pigsties and other small buildings and making enclosures to them which are all certainly becoming a very great nuisance in that situation, & unless now stopped will effectively leave a particularly objectionable encroachment & may hereafter occasion a great deal of trouble to remove, but just xxx may establish rights that may be found exceedingly difficult to displace and for these reasons such external fences, as recommended, should be well defined and no opening or doorway be permitted to be made therein. It will no doubt now be considered a great inconvenience that these encroachments should be required to be removed, but as I have already expressed a favourable opinion that some land should be appropriated for Cottagers' Gardens I consider this a very proper opportunity to require   the railroad Company to purchase, out & out, the whole of this field, No 179, together with all the present encroachments made from it, and the private road adjoining no 180 (both coloured pink) which now appears so especially requisite, so that the company may have full & unrestricted right over the whol , and may then submit under their own arrangements, to · all the accommodations sought for by their tenants, but it should be stipulated the Company shall on being allowed to purchase this ground engage forthwith to erect such a wall as I have described all along on the west side from the turnpike road down to the canal & shown by a double red line on the plan annexed, so as to stop up any footpath being made by the side of the canal, and not either now or hereafter permit any gateway or opening to be made in that wall. The Company at present holds thia field as yearly tenants paying a rent of £27 per annum, but this is a situation, which, if offered for sale by auction in small lots, would sell for a very large price for building upon, and beyond what I propose to advise the trustees to demand from the Company, but under the existing circumstances of the great accommodation which may be desired by the present improper & illegal occupations exercised by usurpation over the premises.

The Company should pay at least the sum of two thousand pounds for a conveyance of this land, as it is freehold and free from land tax.

A similar stone wall should be continued by the Company from the buildings attached to the Royal Engineer public house up to the wall or fence of the church yard & again continued from the south side of those premises belonging to the parsonage, and round the piece of ground now applied for by the Company, being No 140 & 141, until it unites with the fence & the present bank of the railway at A, leaving one high gate at E as a roadway from the Radcliffe Estate as now used there.

The next point that is necessary too remark upon is, as that portion of additional land required by the Company, and coloured brown on the plan annexed hereto, situate immediately adjoining the spouth end of the present station by the side of the present railroad to afford them room for widening the station there it is absolutely necessary to stipulate that the Company shall build a similar wall as before mentioned all along the boundary from B to C and set up a good strong six feet oak pale fence from C to D and from there to Stacey bridges plant and raised to an efficient fence good live quick, and for ever after be subject to keep up and maintain all these fences.

Lands on East Side of Railroad
(coloured brown)

No	Description	State	Quantity	Total Quantity
129	Brook Field	Wheat	4 3 10	

120

128	Pat of "	Clover		
127	" "	Pasture	2 0 9	
126	Paddock	"		6 3 19

Lands on West Side of Railroad
(coloured brown)

140	Brook Field	Barley	2 3 20	
141	G a r d e n s occupied by Workmen		0 2 7	
				3 1 27
		Acres		10 1 6

Subject to the Company undertaking to erect the walls and fences hereinfore stipulated for, I am of opinion the fair and proper sum which the Trustees should require to be paid for conveying to the Company the fee of these lands, containing together 10 1 6, is the sum of two thousand, three hundred pounds (£2,300).

The only remaining point that occurs to me necessary to require the Trustees to stipulate about at this time, is, as to that portion of their property which is situate on the northneast end of the present station, and which I have distinguished in the annexed plan, No 112, by a yellow colour.

A lease has been granted of this, by the Trustees to Messers Clare and Congreve, for 61 years from Lady day 1841, at a ground rent of £27 10 0 (certainly as appears to me not very wisely) but that may not now perhaps be my duty here to observe further upon: however, as this is now occupied and hereafter it may be apprehended it may become more prejudicially appropriated, as to expose the adjoining land, No. 114, called Post Hill Ground, to similar trespasses, as is now experienced on the other side of the station and hereinafter particularly adverted to, that it appears to me the Company would find it much to their advantage, if they could now

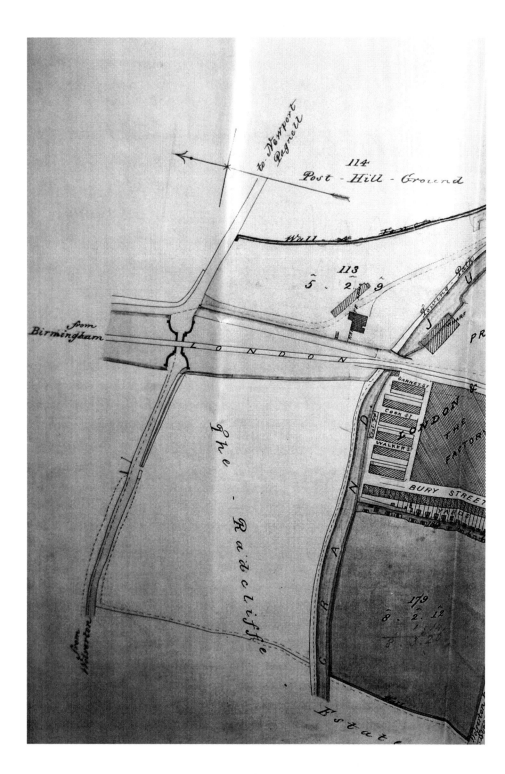

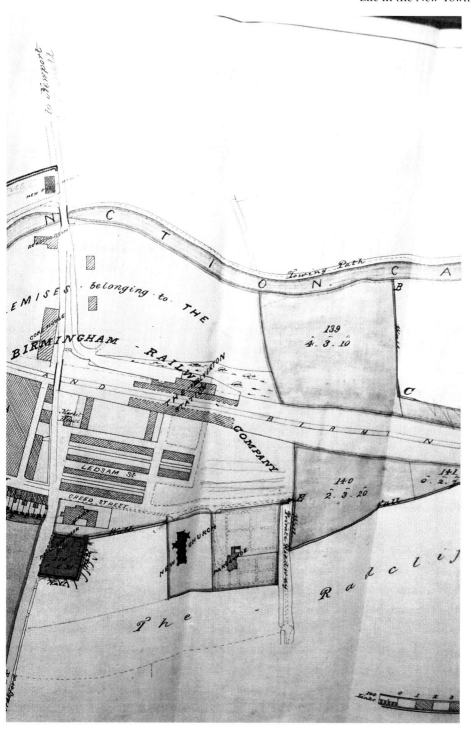

purchase from the Trustees, the fee of all these premises which contain s 5 2 9 as that would give them more influence and control over the lessee; and should the Company be desirous of treating with them for the surrender of such lease, or any portion of the premises comprised in it, which is more than probable they may find it necessary to do, they will be in a better situation, and should the Company determine to avail themselves, at this time, of such opportunity of purchasing the fee; I am of the opinion the Trustees should stipulate for a wall (as before described & shown on the annexed plan by a double red line) being put up along the boundary fence between Nos 113 & 114, which is, according to my opinion, necessary to prevent encroachments & nuisances being committed in the latter field, No. 114, and the trustees should in this case, require the sum of £1,100 for the fee, subject to the subsisting lease, and I see no objection to advising the Trustees to consent to such sale on these terms.

I would presume upon this occasion to direct the attention of the Trustees to the following points referred to me by report, as if these have not been attended to, I would suggest this is the proper opportunity of doing so, the Railroad Company being the party who should defray the expense appertaining thereto.

1st As to the shop gates rendered necessary by the diversion of the river and the protection of the banks and planting quicks to the fence to the river.

2nd As to compensation by the Railway Company for the past use & occupation of Nos. 141, 142 and for damages occasioned thereto & also No. 142.

Probably the Company in making the purchase of the garden ground opposite the Royal Engineer public house may be disposed to purchase these premises also, No. 168 containing 0 1 28 and coloured purple on the annexed plan, as it will give them more power and control over the tenant, and thus regulate proper attention to orderly conduct being observed in it. This is held upon a building lease of 61 years from Lady Day 1841, at a ground rent of £5 0 0 per annum, and although I do not approve of such a selling at all in that situation, I see no objection whatever now in the Trustees selling the fee of it to the Company if they will pay £200, and have it included with the other purchases heretofore alluded to, the same to be inclosed by walls as before set forth, and which are shown in the accompanying plan by a double red line.

Edw Driver
Richmond
April 17, 1847

Mr. Driver attached this plan of Wolverton as it was in 1847. It is hand-coloured but is reproduced in monochrome here. The map was drawn before the extension of Creed, Ledsam and Young Streets and before the later villas were built. The hastily built Radcliffe Arms, isolated by the relocation of the station, is itself planned for relocation to the main road.

The Water Supply

Water borne disease was only just being understood at this time and Wolverton's supply, even though coming from deep wells, was not without its problems. Dr. Corfe used his experience in treating diseases of some railway workers in Wolverton to compile this report.

SANITARY QUESTIONS, OBSERVATION'S, AND SUGGESTIONS ON THE WOLVERTON WELL-WATER, AS SUPPLIED TO THE INHABITANTS.

BY GEORGE CORFE, ESQ. Resident Medical Officer at the Middlesex Hospital.

Soon after the opening of the Railway line from London to Birmingham in 1838, it was considered, by the Directors, that Wolverton, a mere farm-house, fifty-two miles from London, or just halfway between these two towns, should form a station for supplying the locomotives with water, and that a factory should also he established there. The latter has been accomplished, and now the little town averages 1400 to 1600 inhabitants. There are a church, two public-houses, and various shops, so that it has become a place of some importance in every respect.

In order to accomplish the former object, the locomotives were supplied from the canal adjoining the station. The water was forced up by pumps into a tank, and from thence conveyed to the engines when required. But the exorbitant demands which the Canal Company made for such supply, induced the directors of the Railway Company to sink a well in the centre of the factory then erecting. This was in 1840. Accordingly, a well fifteen fathoms deep was sunk, and the water was brought up by an engine to another and larger tank, erected by the side of the former one, but capable of communicating by a main pipe; so that in the event of any deficiency from one source, there might be a supply from the other. As soon as the well, tank, &c., were complete, the engines were wholly supplied by this water; but in the course of a few weeks after its use, the drivers found that they experienced great difficulty in getting the steam up; the water required more fuel, the trains were after time, the boilers and the machinery were much furred and clogged, and the foam or scum on the surface of the boiling liquid was such, that after several failures and disappointments, they were reluctantly compelled to abandon the use of the well-water, and once more return to their supply from the canal.

The houses were now rapidly increasing, and the next question seems to have been, how these dwellings should be furnished with water. The pipes were laid on to the houses, from the well-tank in three or four streets, whilst, in some others, pipes from both tanks were supplied.

It appears, therefore, in the history of this spot, that about the year 1841, the inhabitants were supplied wholly by well-water, and the locomotives by canal water. But it was soon observed that the use of the former beverage gave rise to a singular but unequivocal train of symptoms of derangement in the stomach, bowels, &c. The directors at length were satisfied that the well-water was not only unfit for their engines, but that it was not altogether a wholesome beverage. But then the difficulty arose how these inhabitants should be furnished with a different water. The houses subsequently built were accordingly supplied with "half-and-half," or partly well and partly canal water.

Notwithstanding this arrangement, there was a constant complaint, especially by the new comers, that the water disagreed with them: and the men broke out in eruptions over the face, hands, and neck, and the women and children were the subjects of a much more distressing train of symptoms.

From the year 1841 to the autumn of 1847, a large number of cases were attended by the medical gentlemen of Newport Pagnell, Stoney Stratford, and the practitioner at the station, yet the Company were then in the habit of sending cases occasionally to the Assistant Physician of this hospital: but it appears that about the autumn and winter months of 1847 and 1848, the water, especially of the well, became unusually thick, yellowish, and loaded with extraneous vegetable matter.

Since the 1st of January of the present year, 110 out-patients bave been sent to Middlesex Hospital from Wolverton, exclusive of surgical cases. There has also been a large mortality from pulmonary complaints at the station, the patients, chiefly children, having been attended by the resident medical practitioner.

A large proportion of the out-patients above-mentioned came under my charge as the resident medical officer, on account of their arrival at irregular hours, and I was struck with the uniformity of their symptoms.

Of these 120 cases of which I have carefully preserved an abstract or outline of the diseases, &c., there were thirty-seven males and seventy-three females. Of these males, eighteen were under the age of twelve years, and the remainder were boys or men employed in the factory. Of the females, there were also eighteen under the age of twelve, and the remainder were married women with families, except in two or three instances.

But in order to present the subject in its most important and interesting view, I am desirous of classifying the whole number under two distinct heads, viz., those who partook of the well water *exclusively,* and those who were supplied by the mixed, or canal and well water and the striking difference in the character of the respective diseases will be apparent.

There were sixty-one cases (as the sixty-second patient lived at Stoney-Stratford, and walked to his work daily at the station) who were sufferers from the injurious effects of the well-water, and forty-nine of the mixed water.

My register of the cases specifies the sex, name, age, locality in Wolverton, the period of residence at the station, the state of health before such residence, the towns from whence they came, and the date of their illness, its rise, progress, and their present symptoms, together with the observation of any peculiar features in the disease, and its termination, &c.

The diseases may be divided into two classes. The first class were chiefly amongst the young men, boys, and children, and soon after their arrival in the town they presented the following symptoms: scaly or desquamating surface of the face and neck, pórrigo of the head, or psoriasis of the neck, hands, and arms, or urticaria. Febrile excitement, with more or less disturbance in the bowels, loss of appetite, cough, and disturbed sleep.

Amongst the second class of sufferers from this water, there were the following more serious, distressing, and intractable symptoms:—constant frontal headache, muscse volitantes, dizziness of sight, temporary loss of consciousness, impaired memory, recurring three or four times a day, excessive prostration of the mental and bodily powers, palpitation, pulse varying from 100 to 120, sometimes intermittent, skin feverish, tongue clammy and furred, bowels costive, and constant gastralgia, with tenderness on pressure. No appetite, disturbed sleep from frightful dreams, a dry, harsh, and straining cough. The cerebral symptoms in some instances, were attended with distinct epileptic fits, especially in one lad of seventeen, who was seized with nine successive attacks in one day; but it is important to notice here that the lad lost the attacks of giddiness, &c. T when his uncle, who lives a mile from Wolverton, allowed him to lodge with him, and he walked to the station to his work, but never took his meals there. He expressed himself as not feeling like the same, since he resided at the house of his relation. They all complained more or less of a peculiar coppery or foul taste upon the tongue in the morning, and of the total loss of appetite for breakfast.

Although I have divided the class of patients under two heads, for the sake of drawing a line between the two characters of disease with which they were affected, yet, in reality, the sufferers from the mixed water presented the same train of obstinate and severe com plaints as did those from the well-water, with this difference, that the symptoms in the latter class of patients were more obstinate, more decided, and were never accompanied with the presence of entozoa. Whilst in many of the cases from the canal-water, there were unequivocal evidences of the existence of *ttcnia tata, ascarit,* and *uncarts lumbrimidex,* and after they had been expelled by the usual anthelminties, as turpentine, pomegranate root, Sic , and steel, the symptoms corresponded with those of the patients who partook wholly of the well-water.

In consequence of observing that all these patients exhibited much uniformity in their symptoms, I was induced to note them down, and at length I waited upon Mr. Creed, the Secretary to the Company, at Euston Station, and informed him of my suspicions. This gentleman, after thanking me for the great trouble 1 had taken in investigating the subject, requested me to have an interview with Mr. Dockray, the superintendent engineer. I acceded, and found that he was not altogether a stranger to the suspicious character of the water, but he was under the impression that it merely affected new comers; and that its evil effects soon wore away as the people became accustomed to it, and were able to bear it. In this latter opinion I corrected him, inasmuch as I pointed out the fact that there was a larger number of cases now than had occurred for a long time. However, he was most anxious to obtain all the information possible, and enquired what 1 would suggest. I proposed that the two waters should be analyzed by our Professor of Chemistry, Dr. Ronalds, and 1 left him twenty-four questions, to which he promised to forward replies of undoubted accuracy.

The waters were sent, and the analysis was made, and my questions were faithfully answered, when the following valuable facts came to light. The well-water was proved to be a weak saline and alkaline spring, and the canal-water a tolerably pure soft water.

Here, however, 1 must digress a little in order to return to the subject of the diseases. Until my suspicions were raised that this water was saline and alkaline in its character, I had treated the majority of cases with hydrochlorate of ammonia in some bitter infusion, with an occasional aperient of compound decoction of aloes, or aloetic pills, and by using either the ointment of creosote, the weak nitric oxide of mercury, or chalk, to some of the severe cases of eruptive diseases, but in nearly all the instances, it proved unavailing. When, however, the valuable chemical discovery was made that the well-water was a simple alkaline spring, I immediately commenced a totally opposite treatment. The intro-hydrochloric acid, with or without steel, in some bitter infusion, was now substituted. The eruptive diseases were treated by weak nitric acid lotions, the patients were strictly enjoined not to use the well-water for culinary purposes, and the improvement in their health was very marked and decided, except in two or three instances, where, I afterwards found, I had neglected to warn them not to use the water for the purposes of cooking their meals. But even in these few instances, they made more progress when it was left off than before.

So singular and so uniform were the symptoms of the whole class of patients, that 1 do not exceed the hounds of truth when I assert, that if a patient presented herself, and informed me of the street she lived in, I could repeat her ailments almost word for word before I interrogated her. If she was from Ledsam or Creed Streets, she would labour under the distressing cerebral symptoms already alluded to, with palpitation, hurried pulse, feverish skin, and total prostration of physical and mental energies. If she resided in Walker, Bury, or Cook Streets, or the North Cottages, I might find, in addition to the above symptoms, unequivocal evidences of tsenia or lumbrici, whilst, if they were children, the presence of ringworm, psoriasis, and ascarides, were more or less observed. *One* instance I cannot refrain from alluding to, it was that of a boy eleven years of age, who, with his mother, became patients in the early part of January. They had resided in Bury-street, and had therefore been supplied with the mixed water ; they were natives of Stoney Stratford, two miles distant from the station, and had been in Wolverton eighteen months. The mother was seized in October, 1847, with vertigo, frontal headache, loss of appetite, palpitation, alternate flushings, and clammy perspirations ; and the boy, in November, was attacked with blindness and double strabismus, with constant weight over the forehead. They had both been under medical treatment for some time in Wolverton without avail, and observing the equivocal nature of nervous symptoms in many of these poor people, I was led to conclude that the squint and headache arose from mere cerebral distress rather than cerebral disease, lie was ordered an emetic twice a week, and a dose of turpentine and castor-oil to be given on the morning succeeding the emetic. When he had pursued this plan for a fortnight, the mother and myself were gratified to find that the boy's sight returned, the strabismus diminished, and his health improved ; but it was co-existent with the escape of a large number of joints of a tœnia lata, and numerous ascarides. He was discharged quite well before the end of January, and the mother has greatly improved also under the use of steel wine in a bitter infusion. A man is now in attendance upon Mr. C. De Morgan, who has charge of the Ophthalmic department, with partial loss of sight in the left eye, and impaired vision in the right, attended with músete volitantes. He is a native of Leeds, and had resided in Ledsam Street, Wolverton (well-water), for ten months. This attack crept on after he had been there four months; he is greatly improving.

I will now pass on to the interesting but important facts communicated to me by Mr. R. Dockray, Esq., and Dr. Ronalds. The nature of the earth about Wolverton, for the first fifteen to twenty feet, is a mixture of limestone and white marl, and below that depth it is a blue greasy clay. This clay I have a specimen of, as it was taken from the well, and its consistence, appearance, and chemical qualities, exactly correspond to that of " fuller's earth," so common in the county of Bucks, that is to say, silica, alumina, magnesia, lime, chloride of sodium, with traces of potash and oxide of iron. It will be seen presently that these elements enter largely into the saline ingredients of the well-water. It is forced by a stationary engine into a large cistern in the roof of the factory, and supplies the schools, Creed, Ledsam, and Young Streets, and Glyn Square. The mains are cast iron, whilst the services are partly lead and partly iron pipes.

The locomotives are supplied by the canal, which is also worked into a large cistern on the top of the factory, some distance from the well-cistern, but it is counected with it by a cast-iron main and stopcock. The inhabitants of the North Cottages, Bury and Gas Streets, have a mixed supply of well and canal-water. But these supplies are far inferior in quality to the clear and pleasant springs existing at a public-house, "The Engineer," and the Parsonage House, near the Station. The drainage is excellent, and no suspicion of poison can arise from any imperfection in this department; indeed, the whole train of symptoms in all the cases does not favour the supposition of an animal malaria.

I shall now proceed to the respective analyses of the well and canal water qualitatively, by which the character of this saline alkaline spring will be at once observed.

This water is of a dirty yellow colour, is distinctly alkaline to test paper, it contains chiefly bi-carbonate of lime in solution, and rather more organic matter than should exist in water for drinking.

This water is remarkable for the large quantity of alkaline carbonates, sulphates, and chlorides which it contains, and which render it a weak saline water.

It will be seen by analysis of these two waters, that neither of them are suited for drinking, and that the spring in the Parsonage House is the only one of the two that can be considered to approach to a wholesome beverage.

In addition to this analysis it should be observed that the temperature of the canal is 46°, whilst that of the well is 53°.

I have here subjoined a subsequent analysis made by Dr. Ronalds, of five springs on the station.

The character of the water from the Refreshment Room was any thing but "refreshing" either in look or in taste; it was opalescent, and had a vapid, brackish flavour, which was not observed in the other specimens. The Blue Bridge and Parsonage House waters were very pleasant to drink, and were sparkling in character.

QUALITATIVE ANALYSIS OF SPECIMENS OF WATER FROM FIVE SPRINGS IN AND AROUND WOLVERTON STATION.

No. 1.—Water from the Well in the Refreshment Room.

No. 2—Water from the Blue Bridge Cutting, 1 mile from the station.

No. 3.—Water from the Well at the Parsonage House.

No. 4.—Water from the Well at the Royal Engineer Inn.

No. 5.—Water from the Well at the Radcliffe Arms Inn.

I afterwards found, I had neglected to warn them not to use the water for the purposes of cooking their meals. But even in these few instances, they made more progress when it was left off than before.

Salary Registers

Many of the old London and Birmingham Railway and London and North Western Railway salary registers are preserved in the National Archives. They are interesting sociological documents as well as a valuable resource for descendants of railwaymen.

Three pages pertaining to Wolverton are reproduced here. The clerks, that is the senior people, are well paid, but over a considerable range. Edward Bury is rewarded with a stratospheric £1400 a year, while Joseph Parker, the Works Superintendent, who looked after the day to day running of the plant receives £230. Brabazon Stafford, the chief clerk and accountant was paid £300 a year.

The Engine Drivers are paid by the day. If they worked six days a week they could earn about £100 a year. However, they were not paid if they did not work through illness or accident.

The Police, the men who looked after signalling and safety on the line, were paid 18 shillings on a weekly basis.

REGISTER OF PERMAN

57 [458]   Station, *Locomotive*   Department.

Date of Appoint^t	Name.	No.	Occupation.	Rate of Wages £ s d	£ s d	£ s d	Date of leaving.	Remarks.

Clerks

No Pay day

Wolverton Station, Locomotive Department

Date of Appoint.	Name.	No.	Occupation.	Rate of Wages.						Date of leaving
				£	s	d	£	s	d	

Engine Drivers.

	Name		Occupation							
	Hill W.		Engineman		7 4	7 6	7 5			
					7 10					
✓	Allingh		—		7 6					
	Unsworth J.		—		7 6	7 8	7 10			
					7 4 reduced					
	Ivy W.		—		7 4	7 6	7 5			
					7 10					
	Thompson J.		—		7 2	7 4	7 6			
					7 8					
	Allan James		—		7 2	7 4	7 6			
					7 8					
	Rogers Thos		—		7 4	7 6	7 8			
	Redworth G.		—		7 2	7 4	7 6			
					7 8					
	Penfold A.		—		7 4	7 6	7 8			
	Bennett W.		—		6 8	7	7 2			
	Brown W.		—		7 4					
					6 8	6 16	7 2			
	Shepherd		—		7 4					
					7 2	7 4				
	Holding Thos		—		6 8	7	7 2			
	Sexton Thomas		—		7 4					
					6	6 8				
	Edwards Thos		"		7	7 2				
					6					
	Mullen B.		"		7					
	Guest John		"		7 2	7 4	7 6			
	Blundell Wm		"		7 reduced					
					4 2	5	6			
	Edwards Thos		"		6 8					
					4 2	5	6			
	Brown W.		"		6 8					
					4 2	5	6			
	Hunt Geo		"		4 2	5				

REGISTER OF PERMANENT
Department.

	Date of leaving.	Remarks.

OFFICERS AND SERVANTS.
Wolverton Station, _Police_ Department.

	Name.	No.	Occupation.	Rate of Wages.			Date of leaving.	Remarks.
	Voss William							
	Wallis Edwin							
	Birchall Rich						Aug 1848	To Brighton
	Shaw Edward	554					Oct 19	Dismissed
	Baker William	9					Sep 19	From Oxford
	Crockford Jn	1					Sep 49	To Tring
	Dunham Jn	16					Apr 49	To Northampton
	Wybrow Will							To Wolverton
	Jones Alfred							
	Barham Jas							
	Legg William							
	Scott James	17					Oct 49	To Brighton
	Sherwood Wm	28						
	Matthews John							
	Neal John	29						
	Sidcoak Henry							
	Clarke Rich							
	Shallow James	10						
	Harness Geo	11						
	Smith Joseph	13						

Matters of Scandal and Public Concern

Mr. Blott the Station Master

The job of station master was initially seen as a clerical job. The new station would clearly involve monetary transactions and records would need to be maintained so in 1838 this was a job for a man with training and experience in managing records and keeping accounts and they chose the young Alfred Blott. Mr. Blott was young but the directors obviously thought he would do very well. He had worked at Birmingham since April of that year and had had an opportunity to impress. Those early tickets were written out laboriously by hand and often had to be applied for a few days in advance of travel. Passengers were required to provide their name, address and date of birth, which may illustrate that our desire to collect useless data is not so recent after all. It took some years to develop more mechanized systems of ticket issuance. The key skill was probably identified as clerical and the title of Station Master only emerged after some years of operation. It is also worth bearing in mind that Wolverton was rated a "First Class" station in 1838 so Mr. Blott must have come with glowing credentials. In fact his preferment to this position is a good illustration of how hiring was done in those days. Mr. Blott's father, William, was a farmer in Little Stanmore, on land historically owned by the Duke of Chandos and latterly, through marriage, by the Duke of Buckingham and Chandos. I imagine that William Blott had a word with his landlord who was then able to recommend the young Alfred for this new position at Birmingham and subsequently at Wolverton. In the end, Alfred Blott turned out to be quite competent, although I do not think he was the first choice. This gentleman was G. Kendall who was discharged after a few weeks and Blott took up his position on September 29th 1838, twelve days after the line opened.

He lodged in Old Wolverton until he married in 1843 to a Swiss-born wife Cornelia Vantini. His starting salary was £100 per annum and by 1850 this was £200. I am sure that Mr. Blott would have stayed there and continued to raise his family in Wolverton. He enjoyed a very good income and was a respected member of the community. He was doing very well. However, in 1851 he was at the centre of one of those events that scandalized Victorian Britain - he eloped with a younger woman.

The name of this 'young lady" is not recorded and the affair did not last long for he came back in some disgrace to face up to his discretion. The L & NWR's Road and Traffic Committee sat in judgment on his case and initially decided to move him.

The Manager reported the conduct of Mr. Blott in reference to a late elopement of a young lady from the Wolverton Station - Letters in favour of Mr. Blott were read - also one from himself admitting the truth of the charges against him but stating extenuating circumstances - After full consideration the Committee resolved that Mr. Blott had been guilty of improper conduct and that it be recommended that he be removed from Wolverton and be sent to Oxford as Station Master and further that he be reprimanded by the Manager.

In the meantime there were a number of interventions on Mr. Blott's behalf from some of the worthies in the community, so the matter was reconsidered.

Mr. Blott's case having been again reviewed and the memorial in his favour - his contrition for this offence - together with a letter from Mr. Phillimore on the subject having been carefully considered,

It was resolved

That the notice to Mr. Blott of the directors intention to remove him be withdrawn, and that he be reinstated in his position at Wolverton.

But this was not yet the end of the matter:

Reports were submitted to the Directors of the misconduct of Mr. Bevan, Station Master at Oxford, from which it appeared that he had appropriated excess money belonging to the Company, had been irregular in his habits and had suppressed letters of complaint of his conduct which had been reported to the Directors – Mr. Bruyeres stated that he had investigated the facts, and had every reason to believe that the reports against Mr. Bevan were correct whereupon it was Resolved:

that Mr. Bevan be immediately suspended from duty as an improper person to have charge of the Oxford Station - That Mr. Blott having expressed his willingness to go, be removed to the Oxford station on his present salary - reasonable expenses of removal to be allowed him - That Mr. Shakespeare now at Stamford take Mr. Blott's place at Wolverton at his present Salary, and that Mr. Boor who has for some time past done the Company's business at Stamford take Mr. Shakespeare's place provisionally

So it was finally settled, and it seems that the Company came out of it quite well. They effectively reduced the salary paid to the Wolverton Station Master by £50, replaced an incompetent station master at Oxford with an able one, and removed a stain of social embarrassment from Wolverton. Blott's wife seems to have stuck by him as she was living with him in the 1861 Census. He only spent a few more years at the Railway before becoming Deputy Treasurer of an Oxford College. He died in 1868 at the relatively early age of 50 and his widow moved to

Lewisham to live with her sister-in-law near to one of her sons who practised as an accountant.

Miss Prince and Mr Russell

The school was opened in July in 1841, which is why there is no reference to any schoolteachers in the 1841 Census in either New or Old Wolverton. Until then the residents must have made do without access to education for their children. The first teacher, was a 37 year-old Scot, Archibald Laing. He was appointed by the Board on June 11th 1841 at a salary of £100 per annum and provided with a house and "firing", which we can take to mean a supply of coal. He was still there in 1851. The schoolmaster and his family lived in the school building until the 1890s and I must assume that the living quarters were in the south wing. The two mistresses of record in the 1847 Post Office Directory were Emma Jane Hassall, the girl's mistress and Amelia Prince, who had charge of the infant school. It is possible that they were present during Sir Francis Bond Head's visit. Amelia Prince was the daughter of a London stonemason and probably about 20 years old at the time of Sir Francis's visit. She lodged with Joshua Harris the grocer on Bury Street and for her work was paid only £30 a year. She prevailed, and when Emma Hassall left was promoted to School Mistress with a salary increase to £40 a year. Mr. Laing moved on in the mid 1850s and was replaced by a 25 year- old George Russell and somewhere around all these snippets of fact lies a human interest story.

Mr.. Russell and Miss Prince fell in love, which of course is perfectly fine but for the conventional morality of the day, which did not regard a liaison between two unmarried teachers as proper. Mr.. Russell was dismissed from his post in October 1857 and Miss Prince resigned her position two months later. Through this decision the school lost not only the amorous Mr. Russell but also the very experienced Miss Prince who had worked there for ten years.

George Russell quickly found a job in Poplar so there is no suggestion that he left under a cloud and probably received good references. The future Mrs. Russell, six years his senior in age, joined him in January. They married almost immediately, had one son, and subsequently worked in village schools in Essex and Hampshire.

We can also see the large disparity between men's and women's wages in those times. The schoolmaster's annual salary of £100 put him on a par with clerks and engine drivers but the Infant School Mistress, in this case Amelia Prince, was paid little more than a boy apprentice could earn - typically between £25 and £30 a year. The boy apprentice, on achieving his manhood, could then earn £1 a week or better, whereas Amelia Prince had seen little improvement in her income after ten years

138

of experience. Even so, when the Potterspury Poor Law Union decided to hire a teacher for the workhouse at Yardley Gobion in 1842 they were able to secure the services of a female teacher for only £15 a year.

The Death of a Three Year-Old Boy

from the Bucks Herald, August 2nd. 1851

This is a sad and tragic tale about a woman at the end of her tether who reportedly drowned her three year-old son. Sarah Irons was a cook but in order to gain employment she had to make arrangements for someone to look after her son. Earlier in the year she was working as a cook in Stanmore in Middlesex. Her son was not living with her so one assumes that she had made some arrangement for him to live elsewhere. Seven weeks before this incident she had found work with Wolverton's surgeon, William Rogers, at his house beside the canal. The house was demolished in the 20th century but its former location is marked by the Secret Garden. I have not been able to find a court report about the fate of Sarah Irons, but from the evidence in this report it is highly likely that she was sentenced to death.

NEWPORT PAGNELL, On Monday last, a young woman named Sarah Irons, was brought before the Rev. Geo, Phillimore and W. G. Duncan, Esq., charged with the murder of her illegitimate child, three years of age. John Tyler, constable of Wolverton, deposed that Saturday evening, the 19th July, about half past six o'clock in the morning, he discovered the body of male child in the canal, near to the bridge which leads to the Wolvceton Station from the Newport Road. It had on only a shirt. It had apparently been the water some days. There were no external marks of violence about it. Witness went shortly afterwards to the house of Mr. Rogers, the surgeon, at Wolverton, in whose service the prisoner then was, and questioned her about the child, which she said was at Bradwell. On his telling her that child had been found in the canal, and he had reason to believe it was hers, she said she would tell him all about it, and she then said that on Tuesday morning, she was going to take the child to Bradwell, when it slipped into the canal, and she was afraid to say anything about it. She afterwards made another statement to the effect, that on Tuesday morning she awoke early and found the child nearly dead, and that she took it down the garden and threw it into the canal. Witness took her into custody, and detained her until an inquest was held on the body, when she was released, a verdict of found drowned having been returned by the jury. Ann Harriett Whiffen, a fellow servant of the prisoner's, stated that the prisoner came into Mr. Rogers's service about seven weeks ago—about four weeks ago some one brought child to her at Mr. Rogers. She told witness it was her child. The woman who brought it refused to keep it any longer. The child remained at Mr. Rogers's, with the prisoner, for one night, and the next morning she told witness she had taken the child to Bradwell. A fortnight ago last Saturday, the child was again brought back to Mr. Rogers's to the prisoner. It remained there until the following Monday. It was kept in the prisoner's bed-room, and the door was kept locked. Witness saw the child between five and six o'clock on the Monday evening, it looked thin and pale, but appeared as well as usual. The prisoner went to witness between 11 and 12 o'clock the same night and told her the child was ill. Witness went to the prisoner's room, and found the child dressed and lying on the floor, with its head on a pillow, its mouth open, and it was making a choking noise in its throat. Witness did not stay long, and prisoner said if the child was worse she would call her. She did not do so, and at seven o'clock the next morning, on enquiring about the child, the prisoner said it was better, and she had taken it to Bradwell at five o'clock. On the following Thursday, the prisoner said she had received a letter from a person, and she should not have to pay for the child much longer. She would not tell who the letter came from. On the same evening, she said, it would be happy release for her if the child was to die. She said she had no love for it, as she had none for its father. About half-an-hour afterwards she said, all of a sudden, "it's enough to make one think of doing what they wouldn't do." Witness saw the child, which was found in the canal, and has no doubt it is the child she saw with the prisoner. William Todd, groom in the service of Mr. Rogers, identified the child as being the same lie had seen with Sarah Irons. She told him it was her cousin, and wanted him to take charge of it for 2s. a week, but witness declined doing so. Mr. J. J. Gent, of Stoney Stratford, surgeon, said that he was called in about 12 o'clock on Saturday, the 19th, to examine the body of male child, about three years old, which was stated to have been found in the canal. He made then merely an external examination. The general appearance was healthy. It appeared to have been in the water some days. The mouth was partially open, and the tongue protruding. The pupils of the eyes were much dilated. the following day Mr, of Stoney Stratford, surgeon, made post-mortem examination. On removing the scalp there was red appearance on its internal surface on the left temporal muscle, and immediately above and behind the ear a slight extravasation of blood was perceptible between the integuments and the cranium. The blood vessels were found generally gorged and tended. The vessels of the brain" were turgid, and in that part corresponding with the external appearances extravasation of blood was found, indicating that some injury had been inflicted during life, but witness did not consider it sufficient to cause death. The stomach was free from appearance of inflammation. The heart also was healthy. The lungs were congested, and there was escape of frothy mucus from the nose and mouth. Witness was of opinion that the child did not die natural death. Mr. Freeman confirmed the statement made by Mr. Gent, and was also of opinion that the child did not die a natural death. Witness thought if the child had been thrown into the water after death there would not have been congestion of the lungs or water in the bronchial tubes. Sarah Franklin, Bradwell, gave evidence to the fact of having had the care of the child a few days at the request of the prisoner, who told her it belonged to woman living at the station. John Daniells, a general dealer at Newport Pagnell, said that Thursday, the 17th instant, he bought a child's frock and petticoat and a pair of shoes of the prisoner, at Wolverton. She said they belonged to the child she had asked him to get a home for the week before. Witness asked her where the child was. She said she had sent long way off, as she did not wish to be bothered by the parties who had the care of it. The prisoner, in reply to the charge, said the child died in her bed-room about four o'clock in the morning—that she awoke and found him very silent —he was not breathing, and being alarmed she dressed herself and put the child into a basket, intending to take it to Bradwell, but not being able to do it she went part of the way and turned back, and not having any money or any friends to assist her she was afraid of making an alarm, and not knowing what to do she put it it into the water, but she could say with a clear conscience that the child died a natural death her bed-room. She was committed to take her trial for wilful murder.

Entertainment

Travelling performers had been a feature of European life since at least the Middle Ages so it is no particular surprise to find a group showing up in Wolverton in the 1851 Census, although this is a pure chance discovery. Travelling players could have been in Wolverton on any other week in a ten year period and we would be entirely ignorant. There are, to my knowledge, no surviving playbills or contemporary accounts of such goings on, and local newspapers were yet to be invented in that part of the world. This accidental vignette does show us that there was money to be made in what must have been a hard life on the road. I imagine that after a 56 hour working week, Wolverton's citizens were only too happy to be entertained.

On the night of the census in 1851, both the Radcliffe Arms and the New Inn accommodated the players.

The Rogers family were at the nucleus of this group, spanning three generations: Thomas and Mary Rogers, both 64, their son, also Thomas, with his wife Ann and four children, their daughter Caroline and the man she later married, John Wade Clinton, and two actors in their early twenties, just starting on a career, Charles and Caroline Brown. There was certainly enough of them to form a small acting company, capable of taking on most of the popular dramas of the day. The emphasis was on "light" entertainment and heavy tragedy left to the sophisticates of the London stage. The melodrama was the great favourite. These plays had a plot line which usually boiled down to Dick Dastardly threatening the Virtue of the pale, innocent and defenceless heroine, but thankfully foiled by the manly hero. In addition they might perform short sketches from the Commedia del Arte tradition and do a few comic "turns". They are all recorded in the Census as "Comedians", which would mean that they would perform the repertoire described above rather than do stand-up comedy as we would understand it today. In later censuses the women style themselves as "Actress" and John Wade Clinton gives his profession as "Lecturer and Comedian" which might suggest some changes in their repertoire.

Thomas Rogers the elder was born in Christchurch, Hampshire in 1786. His wife Mary was born in London so it is fair to assume that they met while touring. The family turns up in Warminster in 1841, all of the part of the family business. Their son Thomas is married to Ann with the beginnings of their family. There are three daughters, Amelia, 20, Augusta, 18 and Caroline, 15. Amelia and Augusta disappear from the Census after this date, presumably due to marriage. It is possible they continued their careers.

Some measure of the itinerant lifestyle can be taken from the places of birth of the children of Thomas and Ann Rogers -Agnes in Arlesford, Lavinia in Wimborne, Leonard at Henley in Arden, Amelia at

Christchurch, Alfred at Wimborne, Clara in Somerset. After Caroline married John Wade Clinton, their children were born in Arlesford, Shaftesbury, Stallbridge and Bridport. In every census they are staying at Inns or in lodgings.

Thomas and Mary Rogers probably died with their acting boots on but the next generations appear to move towards more settled professions. Thomas Rogers the younger, his wife Ann, and two of their daughters settle as Innkeepers at Wootton Basset in their 60s. One son, Leonard became a telegraph supervisor in derbyshire and another, Alfred, a bank manager. John Wade Clinton started up a photography business in London's West End. I have not been able to follow Charles and Caroline Brown.

On 30th March 1851, Thomas, the elder and Mary Rogers, John Wade Clinton and Caroline Rogers were staying at the Radcliffe Arms. The New Inn put up Thomas and Ann Rogers and their four children as well as Charles and Caroline Brown. We don't know how long they stayed - I suppose for as many performances as could be booked, possibly a week. I imagine they performed at the Reading Room at Wolverton, this being the only building (apart from the school) able to accommodate this sort of activity.

This is a review from the year of the census and gives a flavour of the pays they were performing and their acting abilities.

The Era: January 19th 1851

Provincial Theatricals

Rugby "Macbeth was performed on Monday to a very full house. Mrs T Rogers, as the lady, was respectable, but there was no indication of her being "top full of direst cruelty". Mr Wade Clinton's Macbeth was distinquished for power and originality. The Macduff of Mr T Rogers was bold and natural. Mr King played Banquo and Hecate and sang the music of the latter very scientifically. The Married Rake followed, in which Miss Caroline Rogers and Mr T Rogers rattled away with spirit and humour"

The Census extracts for 1851 on the following pages show the entries for The Radcliffe Arms and The New Inn, a mile away on the tow path. The actors give their profession as "Comedian" - an older term for an actor which does not carry the present-day connotation.

Census page from 1851 showing residents of The Radcliffe Arms

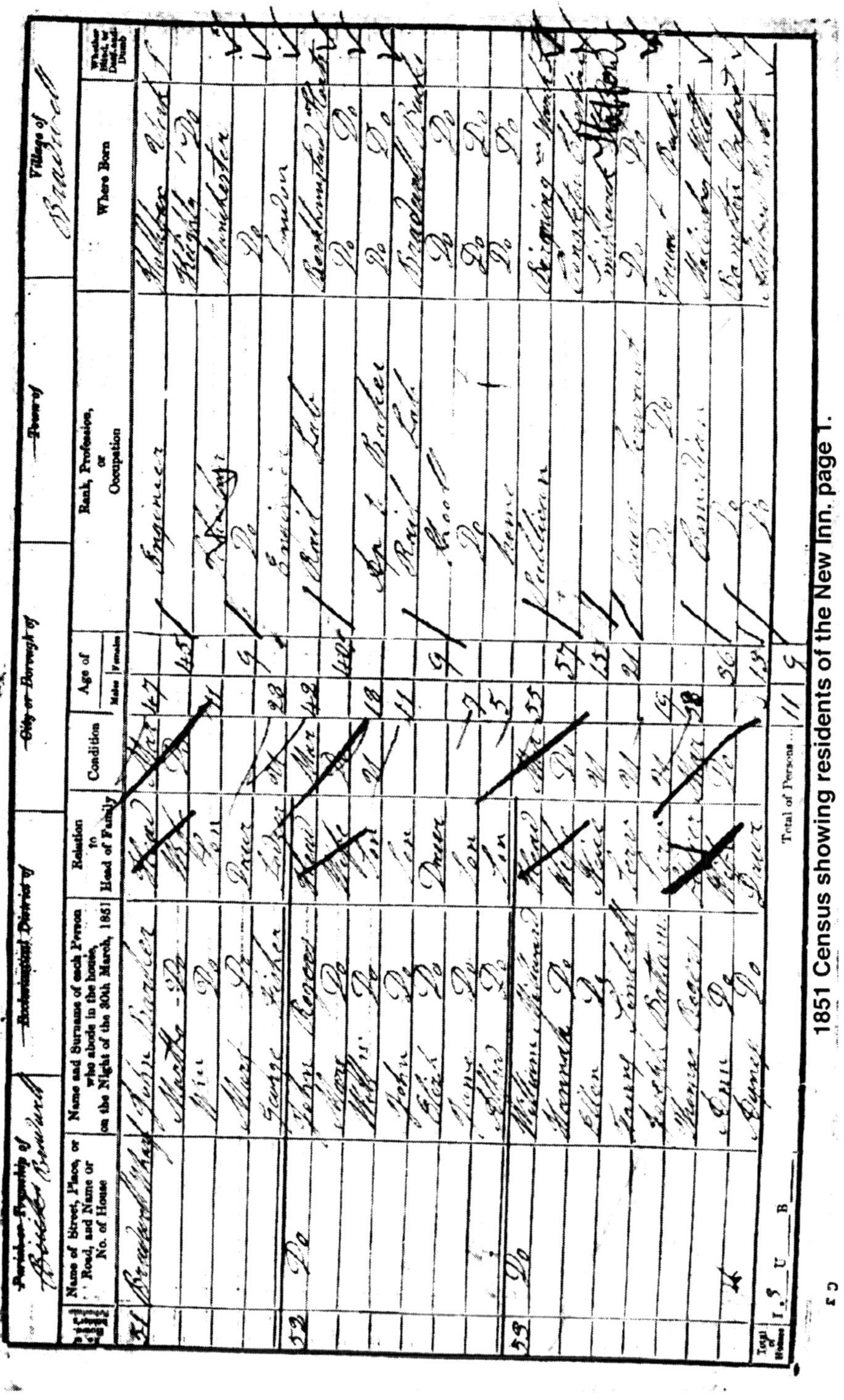

1851 Census showing residents of the New Inn, page 1.

Parish Township of	Ecclesiastical District of	City or Borough of	Town of	Village of

Village of Brancford 272

No. of Householders Houses	Name of Street, Place, or Road, and Name or No. of House	Name and Surname of each Person who abode in the house, on the Night of the 30th March, 1851	Relation to Head of Family	Condition	Age of Males	Age of Females	Rank, Profession, or Occupation	Where Born	Whether Blind, or Deaf-and-Dumb

Total of Persons ... 10 | 10

Total of Houses

1851 Census showing residents of the New Inn. page 2.

146

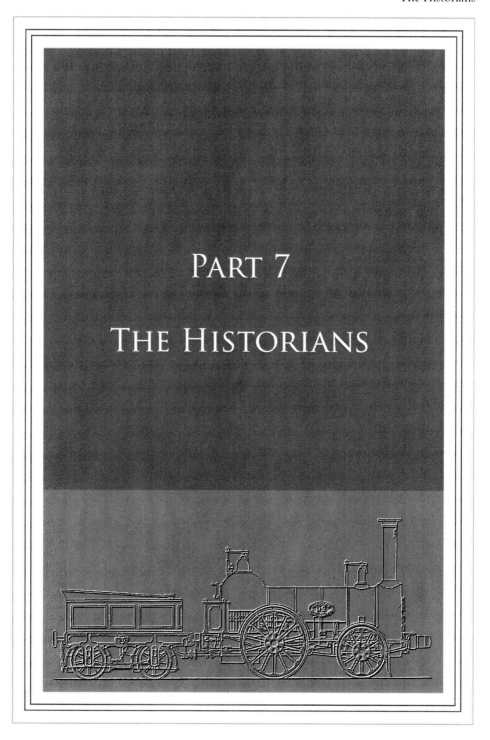

PART 7

THE HISTORIANS

PART 7: THE HISTORIANS

Local history was becoming popular in the 19th century and as we have already seen from the Lyons' account Wolverton and Stony Stratford's history was already being recounted.

The mid-century coincided with the work of two indefatigable recorders of English history and topography. George Lipscomb was writing his four volume history of Buckinghamshire at the very time that Wolverton istef was being created and, although the ancient history of Wolverton is his focus, he took the trouble to record recent developments. In that sense he is a contemporary observer.

James Joseph Sheahan came a little later and published his work in 1862. He makes much reference to the work of Lipscomb and adds his own updated comments.

* George Lipscomb.

The History and Antiquities of the County of Buckinghmam, Volume IV. (1847)

* James Joseph Sheahan.

History and Topography of Buckinghamshire. (1862)

George Lipscomb

From The History and Antiquities of the County off Buckingham
THE LONDON AND BIRMINGHAM RAILWAY enters the County of Buckingham about one mile and a half beyond the Tring Station, and thirty-three and a half miles from London. It traverses the County for twenty-four miles and a half, and passes into the County of Northampton near the small village of Ashton.

This extensive undertaking has its principal station at Wolverton, so far at least as it regards its Locomotive Engines, being situated at a distance of fifty-two miles and a half from the London Terminus, and sixty from that of Birmingham. This gigantic Station may be justly considered one of the wonders of modern times in connexion with Railway enterprises : the land, which a few years ago was covered with rich crops, is now overspread with extensive premises and streets*

The site of this establishment occupies a considerable space: the various buildings and offices arc plain and neat, but fitted up with every convenience, having a frontage on the Grand Junction Canal; and comprise, besides the necessary offices, the Locomotive Engine Depot, the Goods Depot, and the Passengers' Station: the latter is fitted up with suitable waiting-rooms, having female attendants. There arc two elegant refreshment rooms on a very large scale. Here the trains stop ten minutes, for the purpose of allowing passengers to take refreshments; and almost every engine with a train from London or Birmingham is changed at this Station, which answers the double purpose of having it examined, and casing the driver and fireman.

The buildings surround a quadrangle of great extent, the entrance to which is under an archway in the centre of the principal front: besides two side entrances. The Lodge, the Superintendent's Office, and the apartments for business are within the quadrangle, close to the central entrance. In the front of the building are four lines of way, the main double way being in the middle, with an intermediate space of six feet five inches, the whole width of way being about sixty feet.

The erecting shop is on the right of the central gateway, and occupies one half of the front part of this building; it has a line of way down the middle, communicating with a turn table in the principal entrance, and also with the small erecting shop, which is on the left of the gateway. Powerful cranes are fixed in the erecting shops for raising and lowering the engines when required. Contiguous to the small erecting shop, and occupying the chief portion of the left wing, is the repairing shop, which is entered by the left gateway; one line runs down the middle of this shop, with nine turn tables, and as many lines of way at right angles to the central line. This shop will hold thirty-six engines. It is lighted by twenty-four windows, reaching nearly to the roof.

In the same wing, and next to the repairing shop, is the tender wrights' shop, having the central line of way of the repairing shop running down its whole length, with a turn table and cross line, which passes quite across the quadrangle, and intersects a line from the principal entry to the boiler shop in the rear. The remainder of the left wing is occupied by a store-room on the ground-floor, a brass-foundry and store-room over, and the iron-foundry, which extends to the hack line of the buildings. The right wing contains the upper and lower turneries, the upper floor being supported by iron columns. The fixed pump engine-house is also in the right wing, occupying the central portion thereof, in which there are two powerful engines. The smithy occupies the north-west angle of the building, and partly down the right wing. The remaining space of the back portion of the building is occupied by a joiner's shop, with store room and pattern⁀hop, the hooping furnaces, and a boiler-shop.

In the carriage wharf there arc two docks; and the whole width of the landing is about twenty-eight feet, which runs out with a proper slope leading from the station to the main road.

Fronting the canal, and on the east side of the railway, is the goods warehouse, which is furnished with a double way, forming a communication with the main line, with cranes for raising or lowering goods from or to the canal barges, or railway waggons; and beneath, a coal store, with loop-holes next the canal. This building is lighted by four sky-lights in the roof, which is slated, and projects over part of the canal, to protect the barges in bad weather.

WOLVERTON STATION is very compactly and regularly built. It consists of seven streets, a market-house, and nearly two hundred dwellings; the resident inhabitants amount already to nearly 1,000, and as they are rapidly increasing, it is probable that an immediate addition will have to be made to the number of houses. A public road is now (1844) being formed, to connect it directly with the neighbouring town of Stoney Stratford. Several acres of ground are rented by the Railway Company, from the Radcliffe Trustees, which is let in small portions to the Company's servants, for gardens ; and which has been found, in the highest degree, beneficial.

Most ample provision has been made for the intellectual and moral improvement of the Company's servants: there is a large reading-room, well supplied with newspapers,   magazines, &c. and several hundred volumes of books. There are Infants', Girls', and Boys' Schools, which are attended by about 250 children: these schools are under the charge of the Clergyman of the Station, and the expence of conducting them is defrayed by the Railway Company. There is also a weekly evening-school for adults; besides two Sunday-schools, one of which is in connexion with the Established Church, and the other with (lie Wesleyan Methodists, who also have a Chapel here.

NEW CHURCH AT WOLVBRTON STATION. In 1841, the Bishop of (he Diocese, (Lincoln) with the approbation of the Trustees and the Directors, appointed the Rev. George Weight, A.M. to be the resident Chaplain of this great station. This was the *first* appointment of the kind which was ever made. A large room was fitted up, at a great expense, by the Company, which was licenced by the Bishop for (he performance of Divine Service. It will contain 250 people, and is excellently attended. Having been immediately found far too small for the accommodation of the inhabitants, preparations were soon made for erecting a spacious and substantial Church. To meet the expence of this, £1000 were voted by the Company, and another £1000 collected by them; which was paid into the hands of the Radcliffe Trustees, and they, with the most noble generosity, engaged to defray the whole remaining expence, having allotted two acres of ground fur the Church, Churchyard, and Parsonage. The buildings were erected by Messrs. Grissell and Peto, of London ; Messrs. Wyatt and Brandon being the Architects.[7]

The edifice is a very neat structure, in the Early English style of architecture, and consists of a nave, about 80 feet long, by 35 wide, with a spacious chancel at its east end, and which is so constructed as to admit of transepts being afterwards added. The principal window in the chancel is copied from one in Tintern Abbey.[8] There is an ornamental cross on the point of the gable of the roof, which is tiled. At the north-east corner of the nave is a small tower, the eastern side of which, under a Gothic arch, forms one of the principal entrances. The tower is surmounted by an octagonal Spire. The Church is dedicated to St. George the Martyr, and is intended to be made a District Church. The Living is in the gift of the Radcliffe Trustees; and its estimated value about £150 per annum. George Carr Glyn, Esq. Banker, Lombard Street, London, and Chairman of the Railway Company, has presented to the Church a handsome service of communion-plate.

---

7 The first stone was laid 12 July, 1843; and it was probable, that this building, and the adjoining Church Yard, would be consecrated about the month of July; in the present year (1844.)

8 It is intended that thin splendid window shall be ornamented with Stained glass.

James Joseph Sheahan

Extract from his book History and Topgraphy of Buckinghamshire. (1862)

NEW WOLVERTON.—The new town which has sprung up around the Railway Station and the extensive engine works, is now known as New Wolverton; formerly it was called Wolverton Station. It is situated in the parish of Wolverton about 2 miles from Stony Stratford, 1 mile from the parish church, 4 miles from Newport Pagnell, and three-quarters of a mile from the river Ouse.

Here, on the main line of the London and North-Western Railway, 52 1/2 miles from London, is a first class Passengers Station, and adjacent to it, an extensive Goods Depot, having a frontage on the Grand Junction Canal, which intersects this parish, and is carried across the valley in this locality by means of an embankment, and a beautiful aqueduct. The railway viaduct across the same valley is about 600 feet in length, 54 feet high, and has five arches of 59 feet span, and eight smaller arches—four on each side of the larger ones. Here too is the great central Locomotive Engine Depot of the London and North-Western Railway Company, which, together with the Station, covers an area of many acres. Lipscomb, writing in 1844, when the establishment was not near so extensive as it is at present, says:—" This gigantic Station may be justly considered one of the wonders of modern times in connexion with railway enterprises. The land," he continues, "which a few years ago was covered with rich crops, is now overspread with extensive premises and streets." The spacious passengers station is a plain but neat and substantial structure, fitted up with every convenience. The elegant refreshment-rooms are on a large scale, and the waiting-rooms, offices, &c., are complete in every particular.

The very extensive factory for building and repairing the locomotive engines and carriages of the Company, consists of brass and iron foundries, and departments or " shops" for erecting, repairing, and fitting the engines, and for smiths, boiler-makers, wrights, joiners, turners, &c. Powerful cranes are fixed in the erecting shops for raising and lowering the engines when required; and amongst the collection of the most valuable machinery which these works contain, may be noticed the cranes just alluded to, three of the largest sized steam hammers, some very fine lathes, a number of planing, drilling, shaping and slotting machines, and many other beautiful and ingenious contrivances of this kind. The fixed pump engine-house contains two powerful engines. The lodge, the Superintendent's Office, and the apartments for business are all very neat and appropriate. James E. McConnell, Esq., of Wolverton Park, civil engineer, is the General Superintendent of the works. We have seen that the Bishop of Oxford (then the guest of Mr. McConnell) addressed 2,000 workmen in the engine shed of this depot, on the day upon which New Bradwell Church was consecrated; and we shall see anon that the same distinguished prelate addressed a larger assembly in the same place, on the day that the first stone of the Mechanics' Institute was laid.

Many hundred workmen find continual employment here. The number of men and boys employed here at present is between 2,300 and 2,400, and, as has been seen, a large village has been built for their accommodation on the spot, by the Railway Company ; and another village in the neighbourhood, called New Bradwell, owes its origin to the same Company.

New Wolverton, which is very compact and regularly built, consists of several uniform streets, containing about 250 neat houses, replete with every convenience, and supplied with gas and water by the Railway Company. The place contains a Market-place, butchers-shambles, several good shops, two large inns, schools, baths, a dispensary, &c. Several acres of ground are rented by the company, from the Radcliffe Trustees, which is let in small portions to their servants for gardens. In 1844, a public road was formed to connect this place with the neighbouring town of Stony Stratford.

In the year 1841, the then Bishop of the Diocese (Lincoln) appointed the Rev. George Weight to be resident Chaplain in this rising settlement, and licensed a large room for Divine Service, which had been fitted up in an appropriate manner by the Railway Company. This apartment was found to be too small for the purpose, and preparations were soon made for erecting a church. To meet the expense of this the Company voted £1,000, and another sum of £1,000 was collected by them. These two sums were paid into the hands of the Radcliffe Trustees, and they not only engaged to defray the remaining expense, but they allotted two acres of ground for the church, church-yard, and parsonage The church and parsonage were built at a cost of about £5,000. The first stone of the church was laid on July 12th, 1843, and the edifice was consecrated on Whit Tuesday, May 28th, in the year following. By an Order in Council, dated 19th May, 1846, a new Ecclesiastical District was allotted to it, the new district containing then 295 inhabited houses, and a population of 1,666 souls.

The Living is a Perpetual Curacy in the gift of the Radcliffe Trustees, who allow £100 a-year towards the stipend of the minister. The present Incumbent is the Rev. F. W. Harnett.

The Church (St. George the Martyr) is a very neat structure in the Early English style. It consists of a nave and chancel, both spacious and so constructed as to admit of transepts afterwards being added. At the north-east corner of the nave is a small tower, in the eastern side of which, under a recessed arch, is the principal entrance. The tower contains one bell, and finishes with an octagonal spire, covered with lead. The roofs are tiled. The walls are of native stone, with Derbyshire stone dressings. The floors are boarded, and seated with neat open benches ; there is a large gallery at the west end, on which is a good organ enclosed in a handsome case ; the pulpit and reading-desk are of carved wood; and the font, which is circular and sculptured, is placed in a recess or small baptistry on the left of the tower entrance, and in which is a stained glass window. The nave is lighted by fourteen lancet windows, arranged in pairs. The chancel arch is semicircular, and rests upon circular pillars with moulded capitals. The east window, of three lights, is copied from one in Tintern Abbey, and is ornamented with stained glass, representing the Nativity, the Last Supper, and the Crucifixion. The reredos, above the communion-table, is of stone, grained to resemble marble. A handsome service of communion-plate was presented to the church by G. C. Glyn, Esq., banker, London, late Chairman of the Railway Company.

Lipscomb gives plates of the two churches of Wolverton.

The Parsonage House adjoining the churchyard is a handsome Gothic residence of stone. The Schools for boys, girls, and infants, with residences for the teachers, form a somewhat extensive range of red brick buildings, with stone dressings, in the Tudor style of architecture.

About 300 children attend. The Railway Company allow the master a salary of £60 a year with a house and gas; and the mistresses of the girls and infants' schools £20 each.

Whit-Monday, in the present year (May 20th, 1861) will be remembered as a " red-letter day " by the inhabitants of New Wolverton. On that day, chiefly through the efforts of Mr. McConnell (to whom, under the Directors of the Railway Company, and the Lords of the Manor, the place is deeply indebted for much that it enjoys) the foundation stone of a Mechanic's Institute was embedded in the earth. The festivities of the day commenced with Divine Service in the church, when the Lord Bishop of Oxford preached an appropriate sermon to a very crowded congregation, who had gone to the church in procession, headed by a number of " Foresters " from the neighbourhood, and two bands of music. At the termination of the service another procession was formed, and the company repaired to the site of the new building (which is but a short distance from the church) where the " laying " of the stone was performed by his Grace the Duke of Sutherland, who addressed the assemblage in brief but effective terms. The stone being deposited in its bed with the usual formalities, the Bishop delivered a prayer and pronounced a blessing, and the whole assembled multitude struck up the Old Hundredth Psalm in solemn chorus. Immediately afterwards an adjournment was moved to a field where several rural sports came off, and a couple of hours were delightfully spent by the workmen in athletic games and exercises, and by the general company in admiring and applauding their prowess.

The proceedings terminated with a monster tea party in the new engine shed, which had been tastefully laid out and decorated for the occasion. Tables were laid the whole length of the apartment, and an abundant supply of tea, cakes, and other refreshments were dispensed to the working men and their families. A raised platform was devoted to the Bishop of Oxford, the Duke of Sutherland, and other distinguished guests, and from it the Right Rev. prelate addressed between 2,000 and 3,000 persons, in his usual eloquent and effective style. Tea over, the remainder of the evening was spent in the open air amusements which had been interrupted by the tea party.

Among the distinguished personages that witnessed the interesting ceremonies and rejoicings of the occasion, were (besides the Bishop and the Duke already mentioned) the Marquis and Marchioness of Chandos (now Duke and Duchess of Buckingham), the Earl and Countess of Caithness, Earl Ducie, the Earl of Euston, M.P., Sir Harry Verney, Bart., M.P., and many of the clergy and gentry of the neighbourhood.

Deposited in a cavity of the stone, was a box containing some coins of the realm, and a parchment writing, stating that the foundation stone was laid by the Duke of Sutherland, on the above named day, in the presence of the Bishop of Oxford, and that " The building is erected for the instruction and recreation of the workmen employed in the engine establishment of the London and North Western Railway."

The building, which is rapidly progressing towards completion, will be an ornament to the little town, and will contain a lecture-room, concert or music-hall, library, reading-rooms, &c The Railway Directors have granted the site, and the cost of the edifice is expected to be about £1,500. On the day that the erection of the building commenced, 615 workmen had subscribed the handsome stun of £523 towards the building fund. The members of the mechanics or literary institution, possess an excellent library of several hundred volumes, hitherto kept at the schools; one of which has also been used as a reading-room.

PART 8

TRADE AND COMMERCE

In this section

Trade Directories first appeared in the late 18th century, but by the time the new railway town of Wolverton was created they were commonplace. The Robson Directory of 1839 has not had time to catch up to the phenomenon of Wolverton and is not reproduced here. The 1842 Pigot Directory is the first to begin to recognize the new trading community of Wolverton. Five years later Wolverton is a significant community in its own right.

The two reproduced here offer snapshots of the growth of Wolverton.

• 1842 Pigot

Wolverton is still small enough at this date to be included in the larger Stony Stratford directory. Many of the few Wolverton traders were also established in Stony Stratford.

• 1847 Kellys.

Wolverton is now a place in its own right and has a separate entry in this directory.

PART 8: TRADE AND COMMERCE

Pigot and Company's 1842 Dircetory

STONY-STRATFORD, WITH THE VILLAGES OF OLD STRATFORD, WOOLVERTON AND NEIGHBOURHOODS.

STONEY-STRATFORD is a market town, in the united parishes of St. Mary Magdalene and St. Giles, in the hundred of Newport; 51 miles N. w. from London, 14 s. from Northampton, and 2 from the Woolverton station, on the line of the London and Birmingham railway. The town, which is one of considerable antiquity, consists chiefly of a long street, which is gas-lighted; and seated on the Watling-street of the Romans, close to the Ouse, which river here forms the boundary of the counties of Buckingham and Northampton. The parishes are in two manors: the west side being in that of Calverton, the property of the family of Lowndes; the east side in Woolverton manor, belonging to the trustees of the Ratcliff charity. The government of the town is in the resident and county magistrates, assisted by the officers appointed at the manorial courts. Lace is the staple manufacture of the place, and many of the humbler and industrious class of females are employed in that branch — it has, however, greatly declined. Previous to the opening of the great line of railway above mentioned, the town derived considerable advantage as a great thoroughfare.

The parishes had formerly a church each. The church of St. Mary was destroyed by a fire, which occurred in 1742, and consumed with it more than one hundred houses; the tower of the church escaped the flames, but the body has never been restored. The remaining church of St. Giles was erected in 1451, and rebuilt, except the tower, in 1776: the benefice of the united parishes is a perpetual curacy, in the gift of the See of Lincoln: the present incumbent is the Rev. William Henry Bond. There are places of worship for Baptists, Independents, and Wesleyan Methodists. A good national school affords instruction to the children of the poor; a charity of £70. per annum is appropriated to apprenticing children, and there are several other benevolent bequests, from which the poor of the town receive occasional assistance. The market is on Friday; and fairs on the second

Friday in February, the 1st Friday in May, August 2nd, the 3rd Friday in September, the Friday next after 11th October (a statute) and the first Friday In November. The united parishes contained, in 1831, 1,619 inhabitants, and by the last census (1841), 1,757.

OLD STRATFORD is a hamlet, partly in the parishes of COSGROVE, FURTHO, PA8SENHAM and POTTERSPURY, in the county of Northampton and hundred of Cleley, divided from Stonv Stratford by the river Ouse. The hamlet possesses no object deserving notice, and is destitute of all business except that belonging to a small village. Population returned with the parishes. The parish of WOOLVERTON extends to within a mile of Stony-Stratford, but the railway station, which has been the means of bringing Woolverton into notice is, as before-stated, two miles from that town. The parish church of the Holy Trinity is a small structure: the living is a discharged vicarage, in the gift alternately of several persons.   Population about 600.

POST OFFICE, High-street, STONY-STRATFORD, Elizabeth Richardson, Post Mistress. — Letters from London and all parts arrive (by foot post from WOOLVERTON Station) every morning at four, and a second arrival (by the BANBURY Mail) every day at twelve, and are dispatched (by the BANBURY Mail to the WOOVLERTON Station) every morning at a quarter past ten and (by foot post to the same station) every night at twenty minutes before ten.

GENTRY AND CLERGY.

Amos Mrs. Ann, High st

Bond Rev. William Henry, High st

Brooks Miss Elizabeth, Market place

Capes Mrs. Maria, Old Stratford

Chibnall Mrs. Mary, Old Stratford

Clark Mrs. John, High st

Cox Mr. John, Old Stratford

Dickins Mr. John, High st

Durham Mr. John, Old Stratford

Golby Mrs. Mary, High st

Grant Miss Elizabeth, High st

Greaves Mrs. Mary, High st

Harrison Richard, Esq. Woolverton

Hinton Miss Lydia, Woolverton road

Holland Mrs. Mary, Old Stratford

Kipling Mrs. Caroline, High st

Lowndes Mrs. Selby, Whaddon

Mansell Mrs. Henry, Cosgrove

Mansell Mrs. Major, Cosgrove

Oliver John, Esq. High st

Perceval Honourable & Rev. Charles George, Calverton rectory

Popay Mrs. James, Cow fair

Ratciffe Mrs. Ellen, High st

Ratciliffe Mrs. Thomas, High st

Smith Rev. Loraine L., Passenham

Wager Rev. James, Woolverton road

Waite Rev.George, New Woolverton

Wallis Mr. George, High st

Wilkinson Miss Sarah, High st

Yardley the Misses, Priory, Cosgrove

York John, Esq. Market square

ACADEMIES AND SCHOOLS.

Bell Robert (day and boarding), Market square

BRITISH AND FOREIGN, New Woolverton — Archibald Laing, master; Elizabeth Pfeil, mistress

Dodd, Mary Ann, High st [place GentMary (day&boarding),Calverton

Hamblin & Davies (ladies' boarding and day), Market square

INFANTS SCHOOL, New Woolverton - Frances M'Dermott, mistress

Jelley, John (boarding), Belvedere House, Old Stratford

NATIONAL SCHOOL, High St—John Whitamore, master

ATTORNEYS.

Congreve John Freer, High st

Parrott John, High St.

Worley & Kipling, High st

AUCTIONEERS & APPRAISERS.

Durham John (& surveyor), High St.

Harris & Aveline, High st

BAKERS

Corbett John, High St.

Dickins George, Old Stratford

Foulkes Joseph, High St.

Kightly George, New Woolverton

Kightly Henry, Church St.

Paine George, Market square

Paine William, Market square

Paxton James, High st

Truelove Thos.(& confectioner) High st

West Wm. (&confectioner) High st

BANKERS

Bartlett, Parrott & Hearn (branch of Buckingham, open on Fridays), High street – (draw on Praeds, Fane & Co. London)

Olivers & York, High street–(draw on Jones, Loyd & Co. London)

BASKET MAKERS.

Henson Thomas, High st

Smith James, Cow fair

BLACKSMITHS,

Bennett John, High St.

Compton James, Old Stratford

Downing Thomas, Calverton end

Jeffcote Edward, Market square

BOOKSELLERS, STATIONERY AND PRINTERS.

Nixon William (& stamp distributer). High street

Sleath John (& binder), High st

BOOT AND SHOE MAKERS'

Attkins James (warehouse), High st
Barley William, New Woolverton
Benson William, Cow fair
Calladine John, High st
Curtis William, High St.
Foxley William, High St.
French Thomas, Woolverton
Henderson William, High St.
King William, Calverton place
Scaldwell Henry, High St.
Smith Thomas, Old Stratford
Watts William, High st
BRAZIERS AMD TINMEN.
Arnold William, High St.
Revill Edwin, High St.
Sayer George, High st
BUTCHERS.
Gibbins John, High St.
Gilling George, Woolverton
Jarvis Richard, High St.
Richardson Thomas, High St.
Robbins John, Market square
Valentiue Twitchell, High St.
Wilford John, Old Stratford
CABINET MAKERS & UPHOLSTERERS.
Aveline Frederick, High St.
Inwood John, High St.
Pierce Richard, High St.
West George, High st
CARPENTERS & BUILDERS.
Anderson Richard, Horsefair green
Arnold George, High St.
Aveline Frederick, High St.

Day Thomas, High St.

Freeman Thomas, High St.

West George, High st

CHINA, GLASS, be, DEALERS.

Caucutt Mark, High st

Hancock Sarah & Catherine, High st

CHYMISTS AND DRUGGISTS.

Cancntt Mark, High St.

Howe Joseph, Higli st

CLOTHES DEALERS.

Fancutt Frances, High St.

Henderson William, High st

COAL MERCHANTS.

Atterbury William, Old Stratford

Barter Wm.& Richd.Woolverton whf

Freeman Thomas, Old Stratford

Johnson Edward, Old Stratford

COOPERS.

Godfrey Thomas, Market square

Storton John, Market square

CURRIERS AND LEATHER SELLERS.

Harris Richard, High St.

Sharpe William, Church st

FIRE, etc.. OFFICE AGENTS.

ARGUS (life) Joseph Howe, High St.

BRITISH, Thos. Knighton, High St.

CLERICAL, MEDICAL and GENERAL (life) Cattell Hudson, High St.

COUNTY, Wm. Boyes, Market square

NORWICH UNION, William Golby, High street

PHOENIX, Thomas Knighton, High St.

ROYAL EXCHANGE, Robert Bell, Market square

SUN, George West, High st

GROCERS AND DEALERS IN SUNDRIES.

Marked thus • are also Tallow Chandlers.

• Barter Richard, High st

Foulkes Joseph, High st

Harris Joshua, New Woolverton

*Knighton Thomas, High st

Lowe Mary, High st

Nash Thomas, Old Stratford

Parratt James (tea) High st

*Poulter Susannah, High st

Reeve John, High st

Rowlands Elizabeth, New Woolverton

*Scaldwell William. Market square

Shackell William, New Woolverton

Thornton John, High St.

Timbs Martha, Old Stratford

HAIR DRESSER8.

Bonham Thomas, High St.

Brothwell Charles, Market square

Hickson William, High St.

HollowayThomas, High St.

Whichello Ferdinando, New Woolverton

INNS — POSTING.

Bull, Thomas Carter, High St.

Cock, Joseph Clare, High st

IRONMONGERS.

Arnold William, High St.

Day Thomas, High St.

Revill Edwin, High st

LACE MANUFACTURERS.

Cowley Amos, Old Stratford

Loe John, Back st

LINEN & WOOLLEN DRAPERS

Attkins James (woollen) High St.

Attkins James, jun. High St.

Boyes William, Market place

Freshwater Thomas, High St.

King Richard, New Woolverton

Thorne George, High st

MILLERS.

Goodman William, WOOLVERTON MILL

Johnson Edward, STONY STRATFORD MILL-

MILLINERS AND DRESS MAKERS.

Attkins Ann, High St.

Foxley Fanny, High St.

Freeman Sarah, High St.

Harris Mary Ann, Church St.

Wilmin Sarah, Market square

West Isabella, High st

PAINTERS, PLUMBERS AND GLAZIERS.

Collingridge Edward, Church St.

Godfrey George, High St.

Hailey Alfred, Church St.

Parbery John, High St.

Willmer George, High st

SADDLERS AND HARNESS MAKERS.

Sirett William, Church St.

Willison Matthew, Old Stratford*

STONE MASONS.

Hutton John, Calverton place

Richardson Joshua, High St.

Russell William, High St.

Simpson John, Woolverton end

STRAW HAT MAKERS.

Freeman Mary, High St.

Hopkins Sophia, High St.

Jobson Rebecca, High St.

Slade Ann, Old Stratford

SURGEONS.

Back George, High st

Freeman Robert Marriott, High st

Gent John Sleath, Calverton end

Lewis Thos. Harris, Market square

Nixon Daniel, High st

Sutton Robert, New Woolverton

TAILORS.

Attkins James, High St.

Brookes Samuel, Woolverton road

Brown William, New Woolverton

Evans William, High St.

Foddy John, New Woolverton

Godfrey Thomas, High St.

Hooton James, Calverton end

Mackinsay John, New Woolverton

Pater John, High St.

Smith George, Cow fair

Wilson George, High st

TAVERNS & PUBLIC HOUSES.

Angel, John Brown, High St.

Barley Mow, James Bates, High St.

Black Horse,Wm. Atterbury, Old Stratford

Coach & Horses, William French, High St.

Cross Keys, John Parratt (& excise office). High street

Crown, Thomas Blabey, Market square

Falcon, Elizabeth Stockley, Old Stratford

George, William Everett, High St.

King's Head John Robbins, Market square

Locomotive, and Wharf House, Benjamin Barter, Woolverton

Old Royal Oak, Bartholomew Higgins, Calverton place

Plough, William Barter, Woolverton end

Ratcliff Arms, Robert Lambeth Done, New Woolverton

Rising Sun, Thomas Chance, High St.

Shoulder of Mutton, John Key, Calverton

Swan, Mary Hillyer, Old Stratford

Swan, Joseph Marriott, High St.

White Hart, Thomas Prue, Market square

White Horse, Thomas Blabey, High St.

Windmill, Richard Limn, High st

Retailers of Beer.

Blakey Benjamin, New Woolverton

Carter Thomas, New Woolverton

Scaldwell Thomas, High st

Spinks George, New Woolverton

VETERINARY SURGEONS.

Bennett John, High St.

Jeffcoate Richard, Market square

Powell Thomas, Calverton end

WATCH & CLOCK MAKERS.

Arnold William, High St.

Pearson Edmund, High St.

Rust Benjamin, High St.

Tole Henry, High st

WHEELWRIGHTS.

Ganderton John, Woolverton end

Kightly George, High st

WINE &SPIRIT MERCHANTS.

Clare Joseph, High st

Smith Josiah Michael, High st

Miscellaneous.

Abram William, lace pattern drawer, Old Stratford

Anderson Robert, travelling tea dealer,and draper, Woolverton road

Carter Thomas, brewer. High St.

Chibnall Richard, registrar of births and deaths and relieving officer, High St.

Claridge John, parish clerk, High St.

Clark Sarah, seed dealer. Cow fair

Cooke Charles, eating-house, High St.

Frost William, glover. High St.

GAS WORKS. High st—Geo.Sayer,engineer

Golby William, maltster, High St.

Inwood Jane, toy dealer, High St.

Jarvis John, matting manufacturer, Horsefair green

Johnson William, cattle dealer, Old Stratford

Jones John, rope& twine maker. Church St.

Knight John, matting maker. Old Stratford

Lever James, pin maker, High St.

Odell Robt. bricklayer,&c. Market square

Russell William, bricklayer, High St.

Sheppard John, whitesmith, Cow fair

Sunderland Wm. Dyer & scourer, Horn lane

Wilkinson George,brick maker, Brick kilns

COACHES,

To LONDON, the Greyhound (from Birmingham, calls at the George, every morning (Sunday excepted) at four.

To BANBURY, the Royal Mail (from, Woolverton Station) calls at the Cock and the Post Office, every morning at ten minutes past twelve; goes through Buckingham and Brackley.

To BIRMINGHAM, the Greyhound (from London) calls at the George, every night. (Sunday excepted) at twelve.

To the WOOLVERTON STATION, the Royal Mail (from Banbury) calls at the Cock and the Post Office, every morning at ten minutes-past ten.

CONVEYANCE BY RAILWAY.

From the WOOLVERTON STATION, on the LONDON and
BIRMINGHAM Line. From this Station passengers may proceed to almost
any part of England.

An Omnibus, from the Cock, Stony-Stratford, daily, to meet the principal
Trains up and down; and the Banbury Mail calls at the same Inn, on its way
to the Station as above-mentioned.

Particulars of the various Railways are furnished by the Railway Tables.

CARRIERS.

To LONDON, Thomas Golby and - Richardson, from their respective offices,
High street, to the Woolverton Station, and thence by Railway, daily.

To LONDON, Joel Curtis' Waggon, from his house, Church St, every
Tuesday morning; and James Allen, from the White Horse, High street, every
Tuesday night.

To BANBURY, Thos, Golby, daily, – Richardson, every Tuesday & Friday;
and Mary Attkins, from her house, every Thurs, morning, all from High st.

To BRACKLEY, Mary Attkins, from her house, every Thursday.

To LEIGHTON BUZZARD, John Hewitt, from his house every Tuesday.

To NEWPORT PAGNELL, James Russell, from the Coach and Horses, every
Tuesday and Friday evening; and Susannah Claydon, from the Rising Sun,
every Tuesday and Friday forenoon.

To NORTHAMPTON, Mary Attkins and John Holtom, from their houses,
every Wednesday and Saturday.

To TOWCESTER, Thomas Golby and – Richardson, from High street, every
Tuesday and Friday ; and Mary Attkins, from her house, every Tuesday.

Kellys Post Office Directory 1847

WOLVERTON, a parish in the Hundred of Newport, 1 mile east-north-east
from Stony Stratford, contained, in 1841, 1,258.inhabitants, and about 2,081
acres of land; the trustees of the late Dr. Radcliffe are lords of the manor. The
living is a discharged vicarage, in the archdeaconry of Buckingham and diocese of

Oxford, rated in the King's Books at £10 3s. 9d., endowed with £400 private benefaction, and £400 royal bounty; present value about £35; it is in the patronage of the Badcliffe trustees. The church is dedicated to the Holy Trinity ; the Rev. Henry Quartley, M.A., is the incumbent. The London and North Western Railway has a station in this parish, 1 mile from the old church, 2 miles from Stony Stratford. 52 ½ miles on the line from London, and 60 miles from Birmingham; here is the central locomotive depot on the line from London to Birmingham, where the company have erected extensive factories for the repairs, &c, of the locomotive engines which afford employment to a great number of hands; a large number of small houses have recently been built adjacent to the station, occupied principally by the company's officers and servants, which, with a few houses of a superior class, constitute a new town, called Wolverton station. A new district church, with parsonage house, has lately been erected in this part of the parish, at an expense of about £6,000, defrayed principally by the Radcliffe trustees. The living is a perpetual curacy, in the gift of the Radcliffe trustees; the Rev. George Weight, M. A., is the incumbent; present value about £85; the church is dedicated to St. George the Martyr. The railway company have also erected schools, with residence for the master, for children of both sexes, conducted on the British and Foreign system, which are liberally supported, the master's salary being £100 a year, with coals and house; that of the mistress of girls' school, £40; of the mistress of infant's school, £30, with coals and residence; the schools are under the immediate superintendence of the Rev. George Weight, M.A.

Harrison Richard, esq. Wolverton ho
Quartley Rev. Henry, M.A. [vicar]
Weight Rev. George, M.A. [perpetual curate of new church], Station
TRADERS
Arnold John, parish clerk
Assheton Mrs. Sophia, shopkpr. Station
Barter Wm,.' Plough,' & farmer
Barter Benjamin, miller & farmer

Battams William, farmer, Staceys Bushes farm

Blott Alfred, station clerk

French John, shoemaker, Station

Ganderton Wm. & Thos. wheelwrights, agricultural implement makers

Gilling George, butcher, Station

Gostlow Jsph.' Radcliffe Arms,' Statn

Harris Joseph, draper & grocer, Station

Hassell Miss Emma J. schoolmistress, Station

Horwood William, farmer, Park farm

Kightly Geo. baker & confectnr. Station

King Richard, draper & clothier, Statn

Laing Archibald, schoolmaster, Station

Parker Fredk. foreman of locomotive depot, Station

Masters George, Locomotive inn

Prince Miss Amelia, schoolmstrs. Station

Ratliffe Richard, farmer, Stone bridge house farm

Rogers William, surgeon, Station

Salmon Jas. 'Royal Engineer', Station

Shackel Wm. grocer & post office, Station

Simpson Wm, grave stone cutter

Spinks George, coffee & eating ho. Station

Webber Wm. clerk of district church, Station

Wilkinson George Brooks, farmer Manor Farm

Wilkinson George, farmer, Brickhill fm

POST OFFICE.-William Shackel, postmaster. "Letters are received through the office, Stony Stratford, arrive 8 a.m. are despatched 6 even

RAILWAY – to London, Aylesbury, Bedford, Peterborough & Birmingham & places intervening

CANAL - Grand Junction

British Schools (for boys & girls), Archibald Laing, master;

Miss Emma Jane Hassall, mistress

Infant School, Miss Amelia Prince, mistress

COACH to Buckingham, Brackley, & Banbury, daily ½ past 12, from the station.   Omnibus to Stony Stratford on arrival of the trains, 3 times a day from Station

CARRIERS

Richardson, by railway & waggon, to London & Birmingham, dally, & to Stony Stratford & Towcester, tues. & fri. from Station, calls at 'White Horse,' Stony Stratford

To BUCKINGHAM - Lancaster, wed. & fri. from 'Rising Sun,' Stony Stratford; Warner, mon. & thurs. from 'White Hart,' Stony Stratford

To BRACKLEY & BANBURY—Godby, sun. wed. & thurs. calls at the 'Cross Keys,' Stony Stratford

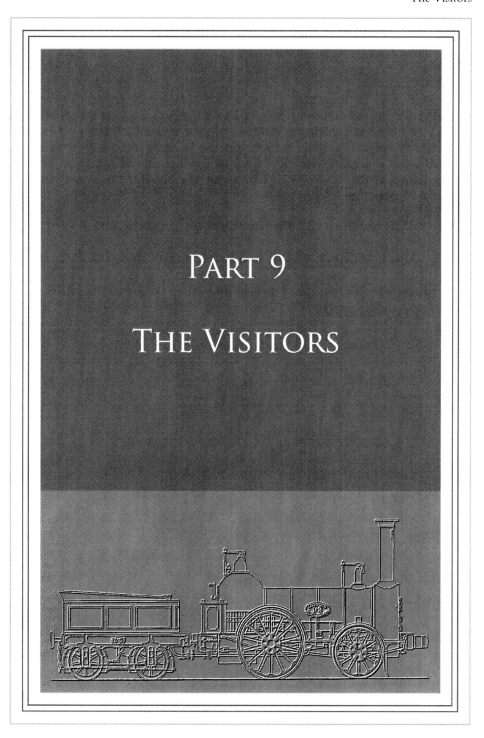

PART 9

THE VISITORS

In this section

Wolverton was famous in the 1840s, not only for its railway workshop, but also for its refreshment rooms as Wolverton Station became for a while the natural stopping place for travellers. It was novel and sensational for the way in which trainloads were efficiently managed. The M1service area at Newport Pagnell gained similar attention when it was created in the late 1950s. Consequently several travel writers took the time to describe Wolverton.

- Hugh Miller, First Impressions of England and its People.

- Queen Victoria's Visit to Stowe, via Wolverton Station

- Letters about the New Road from Wolverton Station to Stony Stratford

- Sir Francis Bond Head, Stokers and Pokers

- Samuel Salt, Railway and Commercial Information

- Samuel Sidney. Rides on railways

- Sir Cusack P. Roney. Rambles on Railways

- Francesca Marton. Attic and Area: The Maidservant's Year

PART 9: THE VISITORS

Hugh Miller. First Impressions of England and its People.

An extract

The earliest traveller was a man called Hugh Miller, a Scottish writer, who, on this occasion, was heading for Olney, where he wished to pay homage to the melancholy 18th century poet, William Cowper. He seems to have arrived on the day of the planned prize fight between two of England's champion bare knuckle fighters, Ben Caunt and Bendigo and therefore finds no room at the raucus Radclffe Arms. Accordingly he walked that night to Newport Pagnell, and failing to find accommodation there, walked a further two miles to Sherington.

I took my seat in the railway train for the station nearest Olney—that of Wolverton. And the night fell ere we had gone over half the way.

I had now had some little experience of railway travelling in England, and a not inadequate idea of the kind of quiet, comfortable-looking people whom I might expect to meet in a second-class carriage. But my fellow-passengers this evening were of a different stamp. They were chiefly, almost exclusively indeed, of the male sex—vulgar, noisy, ruffian-like fellows, full of coarse oaths and dogged asseverations, and singularly redolent of gin; and I was quite glad enough, when the train stopped at the Wolverton station, that I was to get rid of them. At the station, however, they came out en masse. All the other carriages disgorged similar cargoes; and I found myself in the middle of a crowd that represented very unfairly the people of England. It was now nine o'clock. I had intended passing the night in the inn at Wolverton, and then walking on in the morning to Olney, a distance of nine miles; but when I came to the inn, I found it all ablaze with light, and all astir with commotion. Candles glanced in every window; and a thorough Babel of sound—singing, quarrelling, bell-ringing, thumping, stamping, and the clatter of mugs and glasses—issued from every apartment. I turned away from the door, and met, under the lee of a fence which screened him from observation, a rural policeman. "What is all this about?" I asked. "Do you not know?" was the reply. "No; I am quite a stranger here." "Ah, there are many strangers here. But do you not know?" "I have no idea whatever," I reiterated; "I am on my way to Olney, and had

179

intended spending the night here, but would prefer walking on, to passing it in such a house as that." "Oh, beg pardon; I thought you had been one of themselves: Bendigo of Nottingham has challenged Caunt of London to fight for the championship. The battle comes on tomorrow, somewhere hereabouts; and we have got all the blackguards in England, south and north, let loose upon us. If you walk on to Newport Pagnell just four miles—you will no doubt get a bed; but the way is lonely, and there have been already several robberies since nightfall." "I shall take my chance of that," I said. "Ah,—well—your best way, then, is to walk straight forwards, at a smart pace, keeping the middle of the highway, and stopping for no one." I thanked the friendly policeman, and took the road. It was a calm pleasant night; the moon in her first quarter, was setting dim and lightless in the west; and an incipient frost, in the form of a thin film of blue vapour, rested in the lower hollows.

The way was quite lonely enough; nor were the few straggling travellers whom I met of a kind suited to render its solitariness more cheerful. About half-way on, where the road runs between tall hedges, two fellows started out towards me, one from each side of the way. "Is this the road," asked one, "to Newport Pagnell?" "Quite a stranger here," I replied, without slackening my pace; "don't belong to the kingdom even." "No!" said the same fellow, increasing his speed, as if to overtake me; "to what kingdom, then?" "Scotland," I said, turning suddenly round, somewhat afraid of being taken from behind by a bludgeon. The two fellows sheered off in double quick time, the one who had already addressed me, muttering, "More like an Irishman, I think;" and I saw no more of them. I had luckily a brace of loaded pistols about me, and had at the moment a trigger under each fore-finger; and though the ruffians—for such I doubt not they were—could scarcely have been cognizant of the fact, they seemed to have made at least a shrewd approximation towards it. In the autumn of 1842, during the great depression of trade, when the entire country seemed in a state of disorganization, and the law in some of the mining districts failed to protect the lieges, I was engaged in following out a course of geologic exploration in our Lothian Coal Field; and, unwilling to suspend my labours, had got the pistols, to do for myself, if necessary, what the authorities at the time could not do for me. But I had fortunately found no use for them, though I had visited many a lonely hollow and little-frequented water-course—exactly the sort of place in which, a century ago, one would have been apt to raise footpads as one now starts hares; and in crossing the Borders, I had half resolved to leave them behind me. They gave confidence, however, in unknown neighbourhoods, or when travelling alone in the night-time; and so I had brought them with me into England, to support, if necessary, the majesty of the law and the rights of the liege subject, and certainly did not regret this evening that I had.

I entered Newport Pagnell a little after ten o'clock, and found all its inns exactly such scenes of riot and uproar as the inn at Wolverton. There was the same display of glancing lights in the windows, and the same wild hubbub of sound. On I went. A decent mechanic, with a white apron before him, whom I found in the street, assured me there was no chance of getting a bed in Newport Pagnell, but that I might possibly get one at Skirvington, a village on the Olney road, about three miles further on. And so, leaving Newport Pagnell behind me, I set out for Skirvington. It was now wearing late, and I met no more travellers: the little bit of a moon had been down the hill for more than an hour, the fog rime had thickened, and the trees by the wayside loomed through the clouds like giants in dominos. In passing through Skirvington, I had to stoop down and look between me and the sky for sign posts. There were no lights in houses, save here and there in an upper casement; and all was quiet as in a churchyard. By dint of sky-gazing, I discovered an inn, and rapped hard at the door. It was opened by the landlord sans coat and waiscot. There was no bed to be had there, he said; the beds were all occupied by travellers who could get no accommodation in Newport Pagnell; but there was another inn in the place further on, though it wasn't unlikely, as it didn't much business, the family had gone to bed. This was small comfort. I had, however, made up my mind that if I failed in finding entertainment at inn the second, I should address myself to hay-rick the first; but better fortune awaited me. I sighted my way to the other sign-post of the village: the lights within had gone up stairs to the attics; but as I tapped and tapped, one of them came trippingly down; it stood pondering behind the door for half a second, as if in deliberation, and then bolt and bar were withdrawn, and a very pretty young Englishwoman stood in the door-way. "Could I get accommodation there for a night—supper and bed?" There was a hesitating glance at my person, followed by a very welcome "yes;" and thus closed the adventures of the evening. On the following morning I walked on to Olney. It was with some little degree of solicitude that, in a quiet corner by the way, remote from cottages, I tried my pistols to ascertain what sort of a defence I would have made had the worst come to the worst in the encounter of the previous evening. Pop, pop!—they went off beautifully, and sent their bullets through an inch board; and so in all probability I should have succeeded in astonishing the "fancy-men."

Queen Victoria Visits Wolverton

Queen Victoria and her husband were the guests of the Duke of Buckingam at his palatial establishment at Stowe at the beginning of 1845. The journey would take the by railway from Euston to Wolverton and thence by carriage to Stowe. For the railway and North Bucks residents this was a great occasion and great efforts were made. A waiting room was re-decorated for her Majesty and the roads were scraped and levelled. Most of the towns and villages along the route were decorated. This report from the Morning Chronicle details the return journey.

Morning Chronicle January 20 1845

> Her Majesty's Visit to Stowe
> Return of Her majesty
> (From our reporter.)

The principal entertainment provided for her Majesty at Stowe on Friday evening by the care of her noble host, was a concert in which the Messrs. Distins were the performers.

To this concert the invitations were very numerous. The list was given in Saturday's paper.

As the company arrived, something like a drawing-room was held – the guests, on being announced, passing in long array before her majesty, who occupied a throne-like chair in one of the principal apartments.

The Earl of Delawarr and the Duke of Buckingham stood on either side of her Majesty.

During the evening the Queen, observing that some inconvenience was experienced by several of the ladies and gentlemen as they were introduced in approaching sufficiently near to the place she occupied, rose, and herself attempted to move her seat to a more desireable position. The motion was of course anticipated by the watchfulness of her Majesty's attendants, and the position of the chair duly altered.

The concert went off extremely well, her majesty expressing herself as much gratified. The following was the programme:-

Quintet: "Robert toi que j'aime"   Meyerbeer.
Quartet: Prize glee, "Harmony"   Beale
Fantasia: Trumpet, Mr. Distin, "The Soldier Tired," accompanied on the pianoforte by Mr. James Perring   Dr. Arne
Quintet: Etude, "Le Penitent Moir"   Bertini
Quintet: "Fra poco a me" (Lucia)   Donizetti
Quintet: Air de Joseph   Meehul
"God Save the Queen."

About half-past elen o'clock her Majesty and the Prince, attended bythe Duke of Buckingham and the Duchess, passed into the supper room, where they remained for about half an hour.

Shortly after twelve o'clock the Queen and the Prince retired for the night, and the company generally took their departure shortly after one o'clock.

At an early hour on Saturday morning the note of preparation for the departure of the Queen and her Royal Company was sounded.

The portion of the Bucks Yeomanry not selected for escort duty was drawn up near the mansion of Stowe.

The artillery troop took up a favourable position for firing a royal salute.

In Buckingham something like the bustle for the day of arrival was visible. From an early hour the church bells tolled merrily. The flags and banners, which had been kept flying, and the arches and evergreen decorations which had not been removed,looked as fresh and gay as ever. Most of the inhabitants wore ribbons and favours, and the stand erected for spectators was again partially crowded.

Shorty after ten o'clock the royal cortege left Stowe, both her Majesty and the Prince having expressed their delight at the reception they had met with, and their appreciation of the efforts made for their entertainment by their noble host. Bothe The Duke of Buckingham and the Marquess of Chandos rode alongside the royal carriage.

The party passed through the double lines of the yeomanry, the artillery meanwhile saluting, and the band playing the National Anthem.

At Buckingham they were met by townspeople in procession, formed into a somewhat similar order as on the day of arrival.

The usual demonstrations of loyalty and affection were vociferously bestowed on all hands.

After leaving Buckingham, the party proceeded rapidly towards Wolverton.

The escort duty was arranged as before.

At the different arches along the road, groups of the peasantry living in the neighbourhood had assembled, and vociferously cheered the Queen and Prince as they passed by.

At Page-hill the Duke of Buckingham stopped and took leave of his royal guests, returning to Stowe. The Marquess of Chandos accompanied them to Wolverton.

At Stony Stratford, the royal party was met by Lord Carrington, the lord-lieutenant of the county, on horseback. The cavalcade proceeded slowly through the little town, the denizens of which greeted it right loyally. As at Buckingham the evergreens, flags, and ivy still decorated the streets.

The distance from Stony Stratford to Wolverton was soon accomplished, and the cortege drove to the station at a rapid rate.

Inside the station, the staff of the Royal Bucks Militia, and a dismounted party of the yeomanry, under Major Lucas, were drawn up. A number of respectable people had also been admitted to view the arrival and departure of royalty. The usual preparations had been duly made. Crimson cloth was laid over the platform, and the apartment destined for the reception of her Majesty arranged as on the journey down.

Mr Glynn, the chairman of the company, Mr. Creed, the secretary, and several of the principal officials of the railway were in attendance.

The royal party arrived shortly before twelve o'clock.

Her Majesty and the Prince retired for a short time to the apartment provided for them, and then, the special train being reported in readiness, proceeded to the royal carriage. On the platform they took leave of the Marquess of Chandos and Lord Carrington. Prince Albert conversed for some time with the former nobleman, who stood close to the door of the royal carriage.

At twelve o'clock the train was set in motion. Mr. Berry drove the engine. The distance from Wolverton to Euston square, fifty-to miles, was performed in an hour and twenty-five minutes.

On the arrival of the train Mr. Boothby, one of the principal directors was in attendance to receive it, and many ladies were assembled on the platform to greet her Majesty on her return.

The whole of the coachmakers and other mechanics working at the terminus, as well as the servants of the company, were also assembled, amounting in all to between three and four hundred, drawn up on the platform. The assemblage cheered lustily as the train stopped, and her Majesty and the Prince stepped across the platform into the apartment provided for them.

THE RECEPTION-ROOM, AT THE WOLVERTON STATION.

The New Road from Wolverton Station to Stony Stratford

Eighteen months before the visit a new road had been built to join Wolverton Station and Stony Stratford in a direct line. This was not without controversy in 1843 and several were opposed to it. The Queen's visit reignited the controversy as it was noticed that the Queen's party travelled on the old road, brought up to standard for the occasion, rather than take the direct route. We should remember that this was years before the loop line and that the second station was between what is now McConnel drive and the canal. This may help us to understand "A Traveler's" concern about a dangerous turn, obviously shared by the planners of the Royal visit who took a detour that added an extra half-mile to the journey

The Dangerous turn on the new road to the Wolverton Station

To the editor of the Northampton Mercury, Saturday 18th January 1845

Sir: Having been informed by some of the parishioners of Wolverton that the Old Road by Stonebridge House and Mr. Horwood's, in the said parish, has been given up by the Trustees of the Newport and Buckingham Road as a Turnpike, in consequence of the so-called new and improved one going over the railroad, and by the Royal Engineer Inn; will some one of your correspondents do me the favour to inform me if such is the case?

If so, was it by the wish of the said Trustees, or at the suggestion of some kind friend anxious for the safety of our beloved Queen, that she took the Old Road, rather than hazard her personal safety by venturing that most dangerous turn over the railroad through part of New Wolverton on her journey from the station by Stony Stratford to Stowe; thus leaving a Turnpike Road and going into a private one.

I am Sir,

Your obedient Servant,

AN OLD TRAVELLER

Northampton Mercury January 25th 1845

To the editor, Stony Stratford, January 22nd 1845

SIR: "An Old Traveller" is quite correct about the road the Queen went from Wolverton station to this place, viz., not the new turnpike road, about which so much fuss you may recollect was made a year or two ago, but the old, "abandoned" road, which was obliged to be scraped and put in order on purpose for the occasion. So then, it should appear, here we have a Turnpike road recently diverted at great cost,which is so contrived that it is not considered safe to trust the Queen's life upon it, but the old road, no longer turnpike, is preferred on account of its superiority in point of safety, and adopted as the line of her progress towards Stowe.

Yet, in the teeth of this fact, we are to believe, I suppose, that the diverted road is the more "safe and commodious" for all the liege subjects of her Majesty. How a body of trustees could have been got together to vote for this diversion, and then to expend public money in accordance with another vote, in such a preposterous manner, it is difficult to conjecture. But, perhaps, some of your readers could throw light upon this subject; and I think it is simply due to the public in general that it should be explained, otherwise some persons, particularly strangers to the neighbourhood, may run away with a notion that there has been a gross dereliction of duty on the part of a majority of the trustees of the road in question, that is, I presume, the road leading from Buckingham to Newport Pagnel.

I am, sir, your's, &c.,

AN OCCASIONAL READER

Sir Francis Bond Head. Stokers and Pokers.

Extracts

Sir Francis Bond Head, a former Lieutenant Governor of Upper Canada provides the most colourful and detailed account of Wolverton in 1849.

Flying by rail through green fields below Harrow Hill and thence to Watford,—stopping for a moment in a deep cutting to hear a man cry " *Tring !*" and a bell say " *Ring!*" until the passenger gets so confused with the paltry squabble that he scarcely knows which of the two competitors is vociferating the substantive and which the verb,—we will now conduct our readers to the Station and little town of Wolverton.As every city, village, or hamlet on the surface of the globe is usually inhabited by people of peculiar opinions, professions, character, tastes, fashions, follies, whims, and oddities, so there is always to be witnessed a corresponding variety in the allinement and architecture of their dwellings—the forms and excrescences of each often giving to the passing traveller a sort of phrenological insight into the character of the inmates. One street, inhabited by poor people, is as crooked as if it had been traced out by the drunken Irishman who, on being kindly questioned, in a very narrow lane across which he was reeling, as to the length of road he had travelled, replied, " *Faith! it's not so much the length of it as the* BREADTH *of it that has tired me!"* Another—a rich street —'is quite straight. Here is a palace—there are hovels. The hotel is of one shape—the stock-exchange of another. There are private houses of every form—shops of every colour—columns, steeples, fountains, obelisks *ad infinitum.* Conspicuous over one door there is to be seen a golden pestle and mortar—from another boldly projects a barber's pole —a hatchment decorates a third— the Royal Arms a fourth—in short, it would be endless to enumerate the circumstantial evidence which in every direction proves the truth of the old saying, " *Many men, many minds."*

To all general rules, however, there are exceptions; and certainly it would be impossible for our most popular auctioneer, if he wished ever so much to puff off the appearance of Wolverton, to say more of it than that it is a little red-brick town composed of 242 little red-brick houses—all running either this way or that way at right angles—three or four tall red-brick engine-chimneys, a number of very large red-brick workshops, six red houses for officers—one red beer-shop, two red public-houses, and, we are glad to add, a substantial red school-room and a neat stone church, the whole lately built by order of a Railway Board, at a railway station, by a railway contractor, for railway men, railway women, and railway children; in short, the round cast-iron plate over the door of every house, bearing the letters L. N. W. R., is the generic symbol of the town. The population is 1405, of whom 638 are below sixteen years of age; indeed, at Wolverton are to be observed an extraordinary number of young couples, young children, young widows, also a considerable number of men who have lost a finger, hand, arm, or leg. All, however, whether whole or mutilated, look for support to " the Company," and not only their services and their thoughts but their parts of speech are more or less devoted to it:—for instance, the pronoun " *she"* almost invariably alludes to some locomotive engine ; " *he* " to " the chairman," " *it* " to the London Board. At Wolverton the progress of time itself is marked by the hissing of the various arrival and departure trains. The driver's wife, with a sleeping infant at her side, lies watchful in her bed until she has blessed the passing whistle of " the down mail." With equal anxiety her daughter long before daylight listens for the rumbling of " the 3 A.M. goods up," on the tender of which lives the ruddy but smutty-faced young fireman to whom she is engaged. The blacksmith as he plies at his anvil, the turner as he works at his lathe, as well as their children at school, listen with pleasure to certain well-known sounds on the rails which tell them of approaching rest.

The workshops at Wolverton, taken altogether, form, generally speaking, an immense hospital or " Hotel des Invalides " for the sick and wounded locomotive engines of the Southern District. We witnessed sixty of them undergoing various operations, more or less severe, at the same time. Among them was Crampton's new six-wheel engine, the hind wheels of which are eight feet high, weighing thirty-eight tons, and with its tender sixty tons. It is capable of drawing at the usual speed twelve carriages laden with passengers. The workshops at this station are so extensive, that it would be tedious and indeed almost impracticable to describe them in detail; we will therefore merely mention that in one of them we saw working at once by the power of an 18-horse steam-engine twelve turning-lathes, five planing-machines, three slotting-machines, two screw-bolt ditto— and, as a trifling example of the undeviating accuracy with which these contrivances work, we may state that from a turning-lathe a shaving from cold iron will sometimes continue to flow for forty feet without breaking. There are a large cast-iron foundry, a brass foundry, machines for grinding, and also for polishing ; sheers for cutting, and stamps for punching cold iron as if it were pasteboard; an immense oven for heating tires of wheels; a smith's shop containing twenty-four forges, all of which were in operation at once. Two steam-engines—one for machinery, the other for pumping water for the town and offices only, for the Company's well-water here, as at Camden Station, disagrees with the locomotives. A large finishing store, in which were working by steam fifteen turning-lathes, five slotting-machines, five planing ditto, one screwing ditto, two drilling ditto, two shaving ditto. Beneath the above we entered another workshop containing sixteen turning-lathes, two drilling-machines, one slotting ditto, one screwing ditto, one nut ditto, one cylinder-boring ditto, one shaping ditto. In the great store-yard there is an hydraulic press of a power of 200 tons for squeezing wheels on to their axles, or wrenching them off. Another workshop is filled with engines undergoing repair, and adjoining it there is a large store or pharmacopoeia, containing, in the form of oil, tallow, nuts, bars, bolts, &c., all the medicine which sick locomotives occasionally require.

At a short distance towards the south we entered a beautiful building, lighted during the day by plate-glass in the roof, by gas at night, and warmed by steam. In its centre there stands a narrow elevated platform, whereon travels a small locomotive, which brings into the building, and deposits on thirteen sets of rails on each side, twenty-six locomotive engines for examination and repair. On the outside, in the open air, we found at work what is called " *a scrap drum*," which by revolving cleans scraps of old rusty iron, just as a public school improves awkward boys by hardly rubbing them one against another. The scrap iron, after having been by this discipline divested of its rust, is piled on a small wooden board for further schooling, and when sufficiently hot the glowing mass is placed under a steam-hammer alongside, whose blows, each equal to about ten tons, very shortly belabour to " equality and fraternity " the broken bolts, bars, nuts, nails, screw-pins, bits of plate-iron, &c., which are thus economically welded into a solid mass or commonwealth. In another smelting-shop, 150 feet in length, we saw at work fourteen forges, six turning-lathes, one drilling-machine, and one iron shaving machine. Lastly, there are gas-works for supplying the whole of the Company's establishment with about seventy or eighty thousand cubic feet of gas per day.

The above is but a faint outline of the Company's hospital at Wolverton for the repair and maintenance merely of their locomotive engines running between London and Birmingham.

The magnitude of the establishment will best speak for itself; but as our readers, like ourselves, are no doubt tired almost to death of the clanking of anvils—of the whizzing of machinery— of the disagreeable noises created by the cutting, shaving, turning, and planing of iron—of the suffocating fumes in the brass-foundry, in the smelting-houses, in the gas-works—and lastly of the stunning blows of the great steam-hammer—we beg leave to offer them a cup of black tea at the Company's public refreshment-room, in order that, while they are blowing, sipping, and enjoying the beverage, we may briefly explain to them the nature of this beautiful little oasis in the desert.

Wolverton Refreshment-Room.

In dealing with the British nation, it is an axiom among those who have most deeply studied our noble character, that to keep John Bull in beaming good-humour it is absolutely necessary to keep him always *quite full*. The operation is very delicately called " *refreshing him;*" and the London and North-Western Railway Company having, as in duty bound, made due arrangements for affording him, once in about every two hours, this support, their arrangements not only constitute a curious feature in the history of railway management, but the *dramatis 'persona* we are about to introduce form, we think, rather a strange contrast to the bare arms, muscular frames, heated brows, and begrimed faces of the sturdy workmen we have just left.

The refreshment establishment at Wolverton is composed of—

1. A matron or generallissima.
2. Seven very young ladies to wait upon the passengers.
3. Four men and three boys do. do.
4. One man-cook, his kitchen-maid, and his two scullery-maids.
5. Two housemaids.
6. One still-room-maid, employed solely in the liquid duty of making tea and coffee.
7. Two laundry-maids.
8. One baker and one baker's-boy.
9. One garden-boy.

And lastly, what is most significantly described in the books of the establishment—

10. "An odd-man."

"Homo sum, humani nihil a me alienum puto."

There are also eighty-five pigs and piglings, of whom hereafter.

The manner in which the above list of persons, in the routine of their duty, diurnally revolve in "the scrap-drum" of their worthy matron, is as follows:—Very early in the morning—in cold winter long before sunrise—" the odd-man " wakens the two house-maids, to one of whom is intrusted the confidential duty of awakening the seven young ladies exactly at seven o'clock, in order that their "premiere toilette" may be concluded in time for them to receive the passengers of the first train, which reaches Wolverton at 7h. 30m. A.m. From that time until the departure of the passengers by the York Mail train, which arrives opposite to the refreshment-room at about ·eleven o'clock at night, these young persons remain on duty, continually vibrating, at the ringing of a bell, across the rails—(they have a covered passage high above them, but they never use it)—from the North refreshment-room for down passengers to the South refreshment-room constructed for hungry up-ones. By about midnight, after having philosophically divested themselves of the various little bustles of the day, they all are enabled once again to lay their heads on their pillows, with the exception of one, who in her turn, assisted by one man and one boy of the establishment, remains on duty receiving the money, &c. till four in the morning for the up-mail. The young person, however, who in her weekly turn performs this extra task, instead of rising with the others at seven, is allowed to sleep on till noon, when she is expected to take her place behind the long table with the rest.

The scene in the refreshment-room at Wolverton, on the arrival of every train, has so often been witnessed by our readers, that it need hardly be described. As these youthful handmaidens stand in a row behind bright silver urns, silver coffee-pots, silver tea-pots, cups, saucers, cakes, sugar, milk, with other delicacies over which they preside, the confused crowd of passengers simultaneously liberated from the train hurry towards them with a velocity exactly proportionate to their appetites. The hungriest face first enters the door, "magna comitante caterva," followed by a crowd very much resembling in eagerness and joyous independence the rush at the prorogation of Parliament of a certain body following their leader from one house to the bar of what they mysteriously call ' another place.' Considering that the row of young persons have among them all only seven right hands, with but very little fingers at the end of each, it is really astonishing how, with such slender assistance, they can in the short space of a few minutes manage to extend and withdraw them so often—sometimes to give a cup of tea—sometimes to receive half-a-crown, of which they have to return two shillings —then to give an old gentleman a plate of warm soup—then to drop another lump of sugar into his nephew's coffee-cup—then to receive a penny for a bun, and then again threepence for four " lady's fingers." It is their rule as well as their desire never, if they can possibly prevent it, to speak to any one; and although sometimes, when thunder has turned the milk, or the kitchen maid over-peppered the soup, it may occasionally be necessary to soothe the fastidious complaints of some beardless ensign by an infinitesimal appeal to the generous feelings of his nature—we mean, by the hundred-thousandth part of a smile—yet they endeavour on no account ever to exceed that harmless dose. But while they are thus occupied at the centre of the refreshment table, at its two ends, each close to a warm stove, a very plain matter-of-fact business is going on, which consists of the rapid uncorking of, and then emptying into large tumblers, innumerable black bottles of what is not unappropriated called " *Stout*" inasmuch as all the persons who are drinking the dark foaming mixture wear heavy great-coats, with large wrappers round their necks—in fact, are *very* stout. We regret to have to add, that among these thirsty customers are to be seen, quite in the corner, several silently tossing off glasses of brandy, rum, and gin ; and although the refreshment-room of the Wolverton Station is not adapted for a lecture, we cannot help submitting to the managers of the Company, that, considering not only the serious accidents that may occur to individual passengers from intoxication, but the violence and insolence which drunken men may inflict upon travellers of both sexes, whose misfortune it may be to be shut up with them ; considering moreover the ruin which a glass or two of brandy may bring upon a young non-commissioned officer in the army, as also the heavy punishment it may entail upon an old soldier, it would be well for them peremptorily to forbid, at all their refreshment-rooms, the sale by any of their servants, to the public, of ardent spirits.

But the bell is violently calling the passengers to ' Come! come away !' — and as they have all paid their fares, and as the engine is loudly hissing—attracted by their pockets as well as by their engagements, they soon, like the swallows of summer, congregate together and then fly away.

It appears from the books that the annual consumption at the refreshment-rooms averages—

182,500 Banbury Cakes	5, 110 lbs. of moist sugar.
56,940 Queen cakes.	
29,200 pates.	16,425 quarts of milk.

2,920 " coffee.	10,416 " soda-water.
43,800 " meat.	45,012 " stout.
5,110 " currants.	25,692 " ale.
1,217 " tea.	5,208 " ginger-beer.
5,840 " loaf-sugar.	547 " port.

And we regret to add,

666 bottles of gin.
464   ,,   rum.
2,392   „   brandy.

To the eatables are to be added, or driven, the 85 pigs, who after having been from their birth most kindly treated and most luxuriously fed, are impartially promoted, by seniority, one after another, into an infinite number of pork pies.

Having, in the refreshment sketch which we have just concluded, partially detailed, at some length, the duties of the seven young persons at Wolverton, we feel it due to them, as well as to those of our readers who, we perceive, have not yet quite finished their tea, by a very few words to complete their history. It is never considered quite fair to pry into the private conduct of any one who performs his duty to the public with zeal and assiduity. The warrior and the statesman are not always immaculate; and although at the Opera ladies certainly sing very high, and in the ballet kick very high, it is possible that their voices and feet may sometimes reach rather higher than their characters. Considering, then, the difficult duties which our seven young attendants have to perform—considering the temptations to which they are constantly exposed, in offering to the public attentions which are ever to simmer and yet never to boil—it might be expected that our inquiries should considerately go no further than the arrival at 11 P.M. of "the up York mail." The excellent matron, however, who has charge of these young people—who always dine and live at her table—with honest pride declares, that the breath of slander has never ventured to sully the reputation of any of those who have been committed to her charge; and as this testimony is corroborated by persons residing in the neighbourhood and very capable of observation, we cannot take leave of the establishment without expressing our approbation of the good sense and attention with which it is conducted; and while we give credit to the young for the character they have maintained, we hope they will be gratefully sensible of the protection they have received.

Postscript.

We quite forgot to mention that, notwithstanding the everlasting hurry at this establishment, four of the young attendants have managed to make excellent marriages, and are now very well off in the world.

Gardens, Libraries, and Schools.

Before leaving Wolverton Station our readers will no doubt be desirous to ascertain what arrangements, if any, are made by the Company for the comfort, education, and religious instruction of the number of artificers and other servants whom we have lately seen hard at work. On the western boundary of the town we visited 130 plots of ground, containing about 324 square yards each, which are let by the Company at a very trifling rent to those who wish for a garden ; and, accordingly, whenever one of these plots is given up, it is leased to him whose name stands first on the list of applicants. A reading-room and library lighted by gas are also supplied free of charge by the Company. In the latter there are about 700 volumes, which have mostly been given; and the list of papers, &c. in the reading-room was as follows:— Times, Daily News, Bell's Life, Illustrated News, Punch, Weekly Dispatch, Liverpool Albion, Glasgow Post, Railway Record, Airs' Birmingham Gazette, Bentley's Miscellany, Chambers' Information, Chambers' Journal, Chambers' Shilling Volume, Practical Mechanic's Journal, Mechanic's Magazine.

Besides the above there is a flying library of about 600 volumes for the clerks, porters, police, as also for their wives and families, residing at the various stations, consisting of books of all kinds, excepting on politics and on religious controversies. They are despatched to the various stations, carriage free, in nineteen boxes given by the Company, each of which can contain from twenty to fifty volumes.

For the education of the children of the Company's servants, a school-house, which we had much pleasure in visiting, has been constructed on an healthy eminence, surrounded by a small court and garden. In the centre there is a room for girls, who, from nine till five, are instructed by a governess in reading, writing, arithmetic, geography, grammar, history, and needlework. Engaged at these occupations we counted fifty-five clean, healthy faces. In the east wing we found about ninety fine, stout, athletic boys, of various ages, employed in the studies above mentioned (excepting the last), and learning, moreover, mathematics and drawing. One boy we saw solving a quadratic equation—another was engaged with Euclid—others with studying land-surveying, levelling, trigonometry, and one had reached conic sections.

At the western extremity of the building, on entering the infant-school, which is under the superintendence of an intelligent looking young person of about nineteen years of age, we were struck by the regular segments in which the little creatures were standing in groups around a tiny monitor occupying the centre of each chord. We soon, however, detected that this regularity of their attitudes was caused by the insertion in the floor of various chords of hoop iron, the outer rims of which they all touched with their toes. A finer set of little children we have seldom beheld ; but what particularly attracted our attention was three rows of beautiful babies sitting as solemn as judges on three steps one above another, the lowest being a step higher than the floor of the room. They were learning the first hard lesson of this world—namely, to sit still; and certainly the occupation seemed to be particularly well adapted to their outlines ; indeed their pinafores were so round, and their cheeks so red, that altogether they resembled three rows of white dumplings, with a rosy-faced apple on each. The picture was most interesting; and we studied their cheerful features until we almost fancied that we could analyze and distinguish which were little fire-flies—which small stokers—tiny pokers—infant artificers, &c.

On leaving the three rooms full of children, to whom, whatever may be the religion of their parents, the Perpetual Curate, the Rev. G. Weight, is apparently devoting very praiseworthy attention, we proceeded eastward about 100 yards to the church, the property of the Radcliffe Trustees, the interior of which is appropriately fitted up with plain oak-coloured open seats, all alike. In the churchyard, which is of very considerable area, there are, under the north wall, a row of fraternal mounds side by side, with a solitary shrub or a few flowers at the foot of each, showing that those who had there reached their earthly terminus were kindly recollected by a few still travelling on the rails of life. With the exception, however, of the grave of one poor fellow, whose death under amputation, rendered necessary from severe fractures, has been commemorated on a tombstone by his comrades, there exists no interesting epitaph. Besides this church, a room in the library is used, when required, as a Wesleyan Chapel; at which on Sundays there are regular preachers both morning and night—and on Tuesdays and Fridays about 100 of the Company's servants attend extempore prayers by one of their brother artificers.

Samuel Salt, Railway and Commercial Information

Wolverton in 1849.—No. 208.

Upwards of 7,000,000 travellers are annually draughted through Wolverton northward, but they have no opportunity of noticing or knowing, as the trains stop only a few minutes, the rising Railway town, consisting of a series of compact rows of red brick cottages, and forming a complete colony of handicraftsmen and mechanics. A few years ago it was an unmarked spot upon the map, nothing but ploughed and pasture land, bleak, and almost without an inhabitant. Though it stands low and on the banks of the Grand Junction Canal, it is considered very healthy; but it is a remarkable fact, and one that has baffled the inquiries of the sanitarians of the town, that the mortality amongst the children is greater than that of any other town in the kingdom. A sum of £1,500 is disbursed here weekly in wages, and the Company's total stock of engines is 300, which, at £1,500 each, represents a capital invested in locomotives of something like half a million. A sum of £2,500 has been voted by the Directors for the establishment of a mechanic's institution about to be constructed, together with baths and washhouses on the metropolitan principle.

Closing of Capital Accounts.—No. 141. On the 18th February, 1848, Mr. Glyn made the following remarks:—

"I would gladly have closed (and my honourable colleagues concurred with me in that respect), if it were possible, the capital account. It is perfectly easy for a Canal Company or a Dock Company to close their capital account. They buy their land, they finish their works, they seek for trade—and there is, in fact, an end of their capital account at once ; but our case is as different from theirs as circumstances can possibly make it. I remember perfectly well, that at the commencement of this undertaking, under the advice of Mr. R. Stephenson, we purchased of Lord Southampton 22 acres at Camden Town $ and an admirable purchase it was. I remember, however, in my ignorance, inquiring, at the moment, of Mr. Stephenson, why he thought it necessary to buy such an extraordinary quantity of land. To that question he made this reply, which has been impressed on my mind ever since :—' Mark me, Mr. Glyn,' said he,' you require it all; but, if you did not require it, the value of property in that neighbourhood will be so much enhanced, that you would still be able to part with it to advantage.' Now, what has been the result ? We then contemplated to make Camden Town not only our Goods Station and the site of our Locomotive Engine Station for London, but also our Passenger Station. What do we find now? We have removed our Passenger Station down to Euston, and the 22 acres have been found quite insufficient for the purposes of the goods' traffic, and the locomotive necessities of London. We have been obliged to purchase additional land there several times. What have we done at Euston ? We made there what we thought a sufficient purchase from the Duke of Bedford; but since then we have been obliged to buy streets—*streets*, gentlemen—to give to the public the accommodation they require.

And now I tell you that, when these buildings are finished—buildings which, I assure you, are erected without a single ornament, and without the slightest unnecessary expense—when everything is done, we shall not have one foot of ground, or one single room more than is necessary at the present moment. The same observation will apply to Wolverton. We have been obliged to double our capital there; we have been obliged to do nearly the same thing at Crewe. At Manchester, too, we have been obliged to make considerable additions—I speak in the presence of Manchester men as to the necessity of these additions. So, too, at Liverpool, we have been compelled, at last, to increase our Station— for, in fact, the Station there was a disgrace to us. In truth, gentlemen, if, at the commencement, we had doubled the area of all our principal Stations, we should have done right; but we do not,on that account plead guilty to any charge. None of us knew better; we have only derived experience since the commencement of our line ; and the result of that experience is this—that we cannot even now close our capital account. Day after day new requirements are coming upon us for the purposes of our traffic, which is increasing, and will increase; and, so long as that increase goes on, it will be impossible for us, without calling upon yon at once for more capital than it would be fair and right to ask of you, to close our outlay with advantage to yourselves."

Samuel Sidney. Rides on Railways

extract

Samuel Sidney, another seasoned traveller and writer, published this account in 1851.

WOLVERTON STATION.

Wolverton, the first specimen of a railway town built on a plan to order, is the central manufacturing and repairing shop for the locomotives north of Birmingham.

The population entirely consists of men employed in the Company's service, as mechanics, guards, enginemen, stokers, porters, labourers, their wives and children, their superintendents, a clergyman, schoolmasters and schoolmistresses, the ladies engaged on the refreshment establishment, and the tradesmen attracted to Wolverton by the demand of the population.

This railway colony is well worth the attention of those who devote themselves to an investigation of the social condition of the labouring classes.

We have here a body of mechanics of intelligence above average, regularly employed for ten and a-half hours during five days, and for eight hours during the sixth day of the week, well paid, well housed, with schools for their children, a reading-room and mechanics' institution at their

199

disposal, gardens for their leisure hours, and a church and clergyman exclusively devoted to them. When work is ended, Wolverton is a pure republic—equality reigns. There are no rich men or men of station: all are gentlemen. In theory it is the paradise of Louis Blanc, only that, instead of the State, it is a Company which pays and employs the army of workmen. It is true, that during work hours a despotism rules, but it is a mild rule, tempered by customs and privileges. And what are the results of this colony, in which there are none idle, none poor, and few uneducated? Why, in many respects gratifying, in some respects disappointing. The practical reformer will learn more than one useful lesson from a patient investigation of the social state of this great village.

Those who have not been in the habit of mixing with the superior class of English skilled mechanics will be agreeably surprised by the intelligence, information, and educational acquirements of a great number of the workmen here. They will find men labouring for daily wages capable of taking a creditable part in political, literary, and scientific discussion; but at the same time the followers of George Sand, and French preachers of proletarian perfection will not find their notions of the ennobling effects of manual labour realised.

There are exceptions, but as a general rule, after a hard day's work, a man is not inclined for study of any kind, least of all for the investigation of abstract sciences; and thus it is that at Wolverton library, novels are much more in demand than scientific treatises.

In Summer, when walks in the fields are pleasant, and men can work in their gardens, the demand for books of any kind falls off.

Turning from the library to the mechanics' institution, pure science is not found to have many charms for the mechanics of Wolverton. Geological and astronomical lectures are ill attended, while musical entertainments, dissolving views, and dramatic recitations are popular.

It must be confessed that dulness and monotony exercise a very unfavourable influence on this comfortable colony. The people, not being Quakers, are not content without amusement. They receive their appointed wages regularly, so that they have not even the amusement of making and losing money. It would be an excellent thing for the world if the kind, charitable, cold-blooded people of middle age, or with middle-aged heads and hearts, who think that a population may be ruled into an every-day life of alternate work, study, and constitutional walks, without anything warmer than a weak simper from year's end to year's end, would consult the residents of Wolverton and Crewe before planning their next parallelogram.

We commend to amateur actors, who often need an audience, the idea of an occasional trip to Wolverton. The audience would be found indulgent of very indifferent performances.

But to turn from generalities to the specialities for which Wolverton is distinguished, we will walk round the workshops by which a rural parish has been colonised and reduced to a town shape.

* * * * *

WOLVERTON WORKSHOPS.—To attempt a description of the workshops of Wolverton without the aid of diagrams and woodcuts would be a very unsatisfactory task. It is enough to say that they should be visited not only by those who are specially interested in machinery, but by all who would know what mechanical genius, stimulated as it has been to the utmost during the last half century, by the execution of profitable inventions, has been able to effect.

At Wolverton may be seen collected together in companies, each under command of its captains or foremen, in separate workshops, some hundreds of the best handicraftsmen that Europe can produce, all steadily at work, not without noise, yet without confusion. Among them are a few men advanced in life of the old generation; there are men of middle age; young men trained with all the manual advantages of the old generation, and all the book and lecture privileges of the present time; and then there are the rising generation of apprentices—the sons of steam and of railroads. Among all it would be difficult to find a bad-shaped head, or a stupid face— as for a drunkard not one. It was once remarked to us by a gentleman at the head of a great establishment of this kind, that there was something about the labour of skilled workmen in iron that impressed itself upon their countenances, and showed itself in their characters. Something of solidity, of determination, of careful forethought; and really after going over many shops of ironworkers, we are inclined to come to the same opinion. Machinery, while superseding, has created manual labour. In a steam-engine factory, machinery is called upon to do what no amount of manual labour could effect.

To appreciate the extraordinary amount of intellect and mental and manual dexterity daily called into exercise, it would be necessary to have the origin, progress to construction, trial, and amendment of a locomotive engine from the period that the report of the head of the locomotive department in favour of an increase of stock receives the authorization of the board of directors. But such a history would be a book itself. After passing through the drawing-office, where the rough designs of the locomotive engineer are worked out in detail by a staff of draughtsmen, and the carpenters' shop and wood-turners, where the models and cores for castings are prepared, we reach, but do not dwell on the dark lofty hall, where the castings in iron and in brass are made. The casting of a mass of metal of from five to twenty tons on a dark night is a fine sight. The tap being withdrawn the molten liquor spouts forth in an arched fiery continuous stream, casting a red glow on the half-dressed muscular figures busy around, which would afford a subject for an artist great in Turner or Danby-like effects.

But we hasten to the steam-hammer to see scraps of tough iron, the size of a crown-piece, welded into a huge piston, or other instrument requiring the utmost strength. At Wolverton the work is conducted under the supreme command of the Chief Hammerman, a huge-limbed, jolly, good-tempered Vulcan, with half a dozen boy assistants.

The steam-hammer, be it known, is the application of steam to a piston under complete regulation, so that the piston, armed with a hammer, regularly, steadily, perpendicularly descends as desired, either with the force of a hundred tons or with a gentle tap, just sufficient to drive home a tin tack and no more. At a word it stops midway in stroke, and at a word again it descends with a deadly thump. On our visit, an attempt was being made to execute in wrought, what had hitherto always been made in cast iron. Success would effect a great saving in weight. The doors of the furnace were drawn back, and a white glow, unbearable as the noon-day sun, was made visible, long hooked iron poles were thrust in to fish for the prize, and presently a great round mass of metal was poked out to the door of the fiery furnace—a huge roll of glowing iron, larger than it was possible for any one or two men to lift, even had it been cold. By ingenious contrivances it was slipped out upon a small iron truck, dragged to the anvil of the steam-hammer, and under the direction of Vulcan, not without his main strength, lodged upon the block.

During the difficult operation of moving the white-red round ball, it was beautiful to see the rapid disciplined intelligence by which the hammerman, with word or sign, regulated the movements of his young assistants, each armed with an iron lever.

At length the word was given, and thump, thump, like an earthquake the steam-hammer descended, rapidly reducing the red-hot Dutch cheese shape to the flatter proportions of a mighty Double Gloucester, all the while the great smith was turning and twisting it about so that each part should receive its due share of hammering, and that the desired shape should be rapidly attained, sometimes with one hand, sometimes with the other, he interposed a flat poker between the red mass and the hammer, sharing a vibration that was powerful enough to dislocate the shoulder of any lesser man. "Hold," he cried: the elephant-like machine stopped. He took and hauled the great ball into a new position. "Go on," he shouted: the elephant machine went on, and again the red sparks flew as though a thousand Homeric blacksmiths had been striking in unison, until it was time again to thrust the half-welded cheese into the fiery furnace, and again it was dragged forth, and the jolly giant bent, and tugged, and sweated, and commanded,—he did not swear over his task. At length having succeeded in making the unwieldy lump assume an approach to the desired shape, he observed, in a deep, bass, chuckling, triumphant aside, to the engineer who was looking on, "I'm not a very little one, but I think if I was as big again you'd try what I was made of."

Since that day we have learned that the experiment has been completely successful, with a great diminution in the weight and an increase of the strength of an important part of a locomotive.

We have dwelt upon the picture because it combined mechanical with manual dexterity. A hammerman who might sit for one of Homer's blacksmith heroes, and machinery which effects in a few minutes what an army of such hammermen could not do.

If our painters of mythological Vulcans and sprawling Satyrs want to display their powers over flesh and muscle, they may find something real and not vulgar among our iron factories.

203

After seeing the operations of forging or of casting, we may take a walk round the shops of the turners and smiths. In some, Whitworth's beautiful self-acting machines are planing or polishing or boring holes, under charge of an intelligent boy; in others lathes are ranged round the walls, and a double row of vices down the centre of the long rooms. Solid masses of cast or forged metal are carved by the keen powerful lathe tools like so much box-wood, and long shavings of iron and steel sweep off as easily as deal shavings from a carpenter's plane. At the long row of vices the smiths are hammering and filing away with careful dexterity. No mean amount of judgment in addition to the long training needed for acquiring manual skill, is requisite before a man can be admitted into this army of skilled mechanics; for every locomotive contains many hundred pieces, each of which must be fitted as carefully as a watch.

If we fairly contemplate the result of these labours, created by the inventive genius of a line of ingenious men, headed by Watt and Stephenson, these workshops are a more imposing sight than the most brilliant review of disciplined troops. It is not mere strength, dexterity, and obedience, upon which the locomotive builder calculates for the success of his design, but also upon the separate and combined intelligence of his army of mechanics.

Considering that in annually increasing numbers, factories for the building of locomotive, of marine steam-engines, of iron ships, and of various kinds of machinery, are established in different parts of the kingdom, and that hence every year education becomes more needed, more valued, and more extended among this class of mechanics, it is impossible to doubt that the training, mental and moral, obtained in factories like those of Wolverton, Crewe, Derby, Swindon, and other railway shops, and in great private establishments like Whitworth's and Roberts' of Manchester, Maudslay and Field's of London, Ransome and May of Ipswich, Wilson of Leeds, and Stephenson of Newcastle, must produce by imitative inoculation a powerful effect on the national character. The time has passed when the best workmen were the most notorious drunkards; in all skilled trades self-respect has made progress.

A few passenger carriages are occasionally built at Wolverton as experiments. One, the invention of Mr. J. M'Connel, the head of the locomotive department, effects several important improvements. It is a composite carriage of corrugated iron, lined with wood to prevent unpleasant vibration, on six wheels, the centre wheels following the leading wheels round curves by a very ingenious arrangement. This carriage holds sixty second-class passengers and fifteen first-class, beside a guard's brake, which will hold five more; all in one body. The saving in weight amounts to thirty-five per cent. A number of locomotives have lately been built from the designs of the same eminent engineer, to meet the demands of the passenger traffic in excursion trains for July and August, 1851.

It must be understood that although locomotives are built at Wolverton, only a small proportion of the engines used on the line are built by the company, and the chief importance of the factory at Wolverton is as a repairing shop, and school for engine-drivers.

Every engine has a number. When an engine on any part of the lines in connection with Wolverton needs repair, it is forwarded with a printed form, filled up and signed by the superintendent of the station near which the engine has been working. As thus—"Engine 60, axle of driving-wheel out of gauge, fire-box burned out," etc.

This invoice or bill of particulars is copied into a sort of day-book, to be eventually transferred into the account in the ledger, in which No. 60 has a place.

The superintendent next in command under the locomotive engineer-in-chief, places the lame engine in the hands of the foreman who happens to be first disengaged. The foreman sets the workmen he can spare at the needful repairs. When completed, the foreman makes a report, which is entered in the ledger, opposite the number of the engine, stating the repairs done, the men's names who did it, and how many days, hours, and quarters of an hour each man was employed. The engine reported sound is then returned to its station, with a report of the repairs which have been effected. The whole work is completed on the principle of a series of links of responsibility. The engineer-in-chief is answerable to the directors for the efficiency of the locomotives; he examines the book, and depends on his superintendent. The superintendent depends on the foreman to whom the work was entrusted; and, should the work be slurred, must bear the shame, but can turn upon the workmen he selected for the job.

In fact, the whole work of this vast establishment is carried on by dividing the workmen into small companies, under the superintendence of an officer responsible for the quantity and quality of the work of his men.

The history of each engine, from the day of launching, is so kept, that, so long as it remains in use, every separate repair, with its date and the names of the men employed on it, can be traced. Allowing, therefore, for the disadvantage as regards economy of a company, as compared with private individuals, the system at Wolverton is as effective as anything that could well be imagined.

The men employed at Wolverton station in March, 1851, numbered 775, of whom 4 were overlookers, 9 were foremen, 4 draughtsmen, 15 clerks, 32 engine-drivers, 21 firemen, and 119 labourers; the rest were mechanics and apprentices. The weekly wages amounted to £929 11s. 10d.

Of course these men have, for the most part, wives and families, and so with shopkeepers, raise the population of the railway town of Wolverton to about 2,000, inhabiting a series of uniform brick houses, in rectangular streets, about a mile distant from the ancient parish church of Wolverton, and the half-dozen houses constituting the original parish.

For the benefit of this population, the directors have built a church, schools for boys, for girls, and for infants, which are not the least remarkable or interesting parts of this curious town.

The clergyman of the railway church, the Rev. George Waight, M.A., has been resident at Wolverton from the commencement of the railway buildings. His difficulties are great; but he is well satisfied with his success. In railway towns there is only one class, and that so thoroughly independent, that the influence of the clergyman can only rest with his character and talents.

The church is thinly attended in the morning, for hard-working men like to indulge in rest one day in the week; in the evening it is crowded, and the singing far above average.

To the schools we should like to have devoted a whole chapter now, but must reserve an account of one of the most interesting results of railway enterprise.

There is a literary and scientific institution, with a library attached. Scientific lectures and scientific books are very little patronized at Wolverton; astronomy and geology have few students; but there is a steady demand for a great number of novels, voyages, and travels; and musical entertainments are well supported.

The lecture-room is extremely miserable, quite unfit for a good concert, as there is not even a retiring room, but the directors are about to build a better one, and while they are about it, they might as well build a small theatre. Some such amusement is much needed; for want of relaxation in the monotony of a town composed of one class, without any public amusements, the men are driven too often to the pipe and pot, and the women to gossip.

In the summer, the gardens which form a suburb are much resorted to, and the young men go to cricket and football; but still some amusements, in which all the members of every family could join, would improve the moral tone of Wolverton.

Work, wages, churches, schools, libraries, and scientific lectures are not alone enough to satisfy a large population of any kind, certainly not a population of hard-handed workers.

* * * * *

WOLVERTON EMBANKMENT was one of the difficulties in railway making, which at one period interested the public; at present it is not admitted among engineers that there are any difficulties. The ground was a bog, and as fast as earth was tipped in at the top it bulged out at the bottom. When, after great labour, this difficulty had been overcome, part of the embankment, fifty feet in height, which contained alum shale, decomposed, and spontaneous combustion ensued. The amazement of the villagers was great, but finally they came to the conclusion expressed by one of them, in "Dang it, they can't make this here railway arter all, and they've set it o' fire to cheat their creditors."

On leaving Wolverton, before arriving at Roade, a second-class station, after clearing a short cutting, looking westerly, we catch a glimpse of the tower of the church of Grafton, where, according to tradition, Edward IV. married Lady Gray of Groby. The last interview between Henry VIII. and Cardinal Campeggio, relative to his divorce from Catherine of Aragon, took place at the Mansion House of this parish, which was demolished in 1643.

About this spot we enter Northamptonshire, and passing Roade, pause at Blisworth station, where there is a neat little inn.

Rambles on Railways

Strictly speaking, this extract from Rambles on railways is not a "First Impression". Roney wrote and published this almost 30 years after the railways came into being and Wolverton was by this time a mature railway town. In truth Roney adds little to the earlier writers and draws heavily upon Sir Fransic Bond Head's earlier book. I include the extract here largely because it does reinforce the authenticity of the earlier writing. Roney includes facts and statistics which are often wrong and has a tendency to over-write, embellish his supposed learning with Latin tags, and sometimes miss the point altogether. The book reviewer of the Spectator in 1868 sums him up thus:

It is not from want of materials, as we see, that Sir Cusack Roney suffers ; but from an intolerable inability to write, and a still more hopeless inability to strike out half of what he has written.

In spite of this, there are some things in here for the student of Wolverton's history.

Extract from Rambles on Railways

Time was when Wolverton was looked upon as the most dangerous part of the whole line between London and Liverpool. It was, therefore, a halting place not only for all passenger trains, but goods trains came to a stand there also. At Wolverton likewise were the main locomotive repairing shops of the original London and Birmingham Company; and when the company was amalgamated with the Grand Junction and other companies, the locomotive establishment became that for the southern division of the London and North-Western Company, the two northern points of which are, Birmingham, looking westward, and Stafford, looking slightly to the eastward. In process of time the repairing shops have been doubled, trebled, nearly quadrupled. In 1840 Wolverton had a population of 2,000, all of whom were company's servants or their families. In 1849 it was double that number, and now, in 1867, it is 6,000. But of this number about 1,700 are on the wing with their families. These are the men who have hitherto been employed in the locomotive works, which (as stated at page 199) have just been transferred to Crewe. These men and their families, however, will not leave a void in the population, as their places will be supplied from the carriage portion of the Crewe establishment, and from the carriage works of the company, until now located at Saltley, near Birmingham.

The chief owners of land at Wolverton are the trustees of the Radcliffe Library Estate at Oxford; and, although they have erected upon the property a church, to which is attached a churchyard (already beginning to show a great many mounds), they have always been unwilling to dispose of land for building purposes to the extent required by the company. The consequence has been that part of the town or village of Wolverton has had to be built nearly a mile away from it! – at a place called Stantonbury, to the great inconvenience and discomfort of all the men who have to take up their quarters in that locality.

Besides the Radcliffe Episcopalian Church, there is one built mainly by subscriptions from the shareholders of the company. Conjointly they are capable of seating about a thousand people, and the schools connected with them have nearly 600 children in daily attendance. Besides these two churches, there are other places of worship, equal to the accommodation of about 1,100 people.

Of the Infant School Sir Francis Head gives a description, which is as accurate for now as it was for eighteen years ago: – 'At the western extremity of the building, on entering the infant-school, which is under the superintendence of an intelligent-looking young person of about nineteen years of age, we were struck by the regular segments in which the little creatures were standing in groups around a tiny monitor occupying the centre of each chord. We soon, however, detected that this regularity of their attitudes was caused by the insertion in the floor of various chords of hoop iron, the outer rims of which they all touched with their toes. A finer set of children we have seldom beheld; but what particularly attracted our attention was three rows of beautiful babies, sitting as solemn as judges on three steps one above another, the lowest being a step higher than the floor of the room. They were learning the first hard lesson of this world, namely, to sit still; and certainly the occupation seemed to be particularly well adapted to their outlines; indeed, their pinafores were so round, and their cheeks so red, that altogether they resembled three rows of white dumplings, with a rosy-faced apple on each. The picture was most interesting; and we studied their cheerful features until we almost fancied that we could analyse and distinguish which were little fire-flies, which small stokers,which tiny pokers, infant artificers, &c."

209

In the early days of Wolverton, a reading-room, and a library, containing some 700 volumes, supplied the mental occupation and recreation then afforded to its residents. Now there is a Science and Art Institute, an off-shoot from South Kensington, which contributed £500 towards the erection of the building. Its library possesses nearly 3,000 volumes, almost all of which are contributions. The chief person in this work of kindness and goodness is Miss Burdett Coutts; that pre-eminently good lady whose name has but to be mentioned to ensure for it universal respect and admiration.

The institute has an enrolment of 350 members, a large number of whom attend the evening class, and it is a pleasurable fact to state that the pupils have been successful,more than on the average, in the science examinations which are annually held there. As at Crewe, the Government makes it free grants of patent specifications; many of these are closely studied, and all are highly appreciated by the students.

The directors and principal officers of the company, always mindful of the best interests of its staff, and ever keeping a paternal, but not obtrusive eye upon it, have recently erected a model lodging-house, solely for the convenience of single young men. Fifty of them are now accommodated in it; each has a separate bed-room, and the whole establishment is superintended by a very carefully selected manager, who is responsible, not only for the good conduct of the lodgers whilst under its roof, but also for their comfort. The system works well, and it will be extended.

In 1840, and for some few years afterwards, passengers ran a risk at Wolverton to which, happily, or, as we venture to thinks very unhappily, they are no longer exposed — ,that is passengers who travelled chiefly in first-class carriages, and in the express trains of that period were accustomed to alight for ten railway minutes {anglice five) at the celebrated refreshment rooms, the fame of which was world-wide. It was not only that the soup was hot, and the coffee "super-heated' but it was admitted by those who, by the process of blowing the former, and pouring the latter into saucers, were able to get a mouthful or two, it was admitted we say, that each of these beverages was excellent. But there was an attraction at these refreshment rooms that rose superior to all the hot soup, the hot coffee, the hot tea, the buns, the Banbury cakes, the pork pies, the brandy, whiskey, gin, and "rich compounds," the ample statistics of which will be found in our foot-note.[9] Need we say that we refer to the charming young ladies, in whom were concentrated all the beauty and grace that should be corporated in modern Hebes. Our excellent friends, Messrs. Spiers & Pond, of well-earned and well-deserved ,'Buflfet', celebrity, have worthily followed in the footsteps of the great inventress of the railway refreshment room, as it should be; and happy we are to record the fact, that go where we may, we are sure to see, under the magic words announcing that they are caterers, sweet faces, worthy types of English beauty, all the more worthy because with them is combined the modest demeanor emblem of purity, without which all is absent that adorns woman and renders her enchanting. Messrs. Spiers & Pond have — as the late Mrs. Hibbert had — ,but one rule for 'tainted angels ,' — their expulsion.

At moments, however — they were only moments — female grace was at fault at the Wolyerton refireshment rooms. The late Douglas Jerrold (father of one of the workman's most real and truest friends — ,Blanchard by Christian name) had in his play of the *"Housekeeper,"* one of the characters, a drunken wine-porter, who appears on the scene for only a few minutes, and all the language he gives utterance to is advice to his daughter and her companion, never to go any-where without a cork-screw. No doubt this is good paternal advice, such as any good father of a family might give, but it is the use only of the instrument with which we are concerned. Wolverton had many stringent rules, and one of them was that "draught bitter*' should not be "drunk on the premises"; pale ale, therefore, could only be furnished by means of the cork-screw. Now, we appeal to any father, husband, brother, cousin, or lover (the two latter often synonymous, — see all the dictionaries, classical and vulgate); Did you ever see a young lady draw, with grace, a cork out of a bottle in the old-fashioned way, that is, by placing it within the ample folds of her dress (all the more ample if crinoline were concealed behind it), and then tugging until the cork is extracted; if the cork be an easy, obedient, willing cork, the operation is not difficult, and woman's want of grace is but for the moment, but it became momentous, to say nothing of bursting of tapes and wrenching of hooks

---

9 As a footnote here, Roney reproduces verbatim several pages of head's description of the Refreshment Rooms. Since the original is reproduced elsewhere in this book, it is omitted here.

and eyes, red face and perhaps disappointment, if main force must be resorted to. At all events, the late Mrs. Hibbert (known at Wolverton and elsewhere as Generalissima) appreciated the difficulty, and with woman's tact transferred, by a wave of her sovereign sceptre, the beer bottle drawing department of the establishment to the hands of the young gentlemen with all the buttons, and thus released the young ladies from the duty. She never, however, could, to the day of her death, make up her mind that the young ladies ought to be relieved from ginger beer and soda water.

But before we quit for ever (scriptorially) the subject of bitter beer — of Bass, Ind, and Allsopp ; of immortal Burton, that squeezes quart bottles into pints, pints into thimblefuls, of which three-fourths are froth ; and of tap-tub measurement that, by a talisman, converts an imperial pint of the amber fluid into four half-pint *glasses*, let us ask permission to philosophise for a moment — ,for a moment only. Woman! You are never more charming, more feminine, more enchanting than when you are domestic. A magic circle of fascination then surrounds you. You are in your real mission, and being real, you are angelic. But, woman, be true to yourself; be domestic to the fullest extent that brightest imagination can picture or truth realise. But, sex most dear, most loveable of all things human that can be loved, hear the advice of one who believes you were sent on earth for the holy purpose of refining man, and of purifying him — ,never, oh, never be seen using a cork-screw!

Sir Francis Head, in a passage which we purposely omit because we want to have our own say, in our own way, on the subject, informs us that by 1849 four of the young ladies had managed to make excellent marriages. Sir Francis has greatly understated the number. It is quite true that the daily occupations of the young ladies, even without drawing the corks of beer bottles, were arduous and unceasing. Nevertheless, as with all busily occupied people, a time can be found for everything. Not four, but four times four of them found sixteen eligible husbands, and at the present time we know two of them, one not fat, but " fair and forty," the other with slight disadvantage in point of age - forty-four (she confesses to forty[10],) —,

---

10 Ladies, it is sometimes dangerous to conceal your exact ages. We 'will give you a case in point, that only occurred in the summer of the present year. A lady, as far back as 1826, insured her life for the benefit of her relatives. She only died a few months ago; but on coming to compare her age, as given by herself at the time of effecting the insurance, with

but in every other respect at least as eligible, who have had each to exhibit the sable signs of sorrow, void, and bereavement, within the last eighteen months. Let us just pause for a moment, to shed a "pensive tear" to the memory of the two dear departed, just as in the days of our boyhood our sympathies were requested in memory of the celebrated bonnie lassie of "Kelvin Grove," by the father of a lady of present times, who is worshipped by millions, and has been possessed of only by four. (May he of the strong shield endure for ever I) Our tear is shed; and now, like the military bands that accompany the remains of a departed comrade to the grave with the *Dead March in Saul,* and return to barracks with joyous and festive music, do we proclaim, by sound of wedding trumpet and comet-a-piston, the probability that, ere long, each of the charming widows will make a second matrimonial venture* We can, in fact, go one step farther. One of the ladies has already purchased the grey silk dress, absolutely necessary on such occasion ; the second has not gone so far as actual purchase, but she knows where to put her hand upon one at a moment's notice. The dress-maker has already been consulted about the trimmings.[11]

It may probably be observed by any person who has been so venturous as to read the first hundred or so of our pages, that we are given to statistics. This is so, and it has also been alleged of us that we readily detect errors in them when prepared by others. Without taking to ourselves move than a decorous quantity of flattering unction, we believe we shall be able to show, in a work preparing for early publication, that, as regards Post Office statistics, at all events, none have been issued by the department for the last fifteen years that are not abounding in most egregious blunders, and that the logical dogmas of "contrariety, sub-contrariety, and contradiction', were never carried to greater extent than in these documents. Our statistics and contradictions are, at the present moment, however, of a different character. They refer to what, in the palmy days of Wolverton as a

---

that on the certificate of births required by the office, to be obtained, after death, from the parish register, it was found that although the lady was in reality 42 years, in 1825, she only owned to 35, and paid premiums on that scale for 42 years. The office, had it been so disposed, might have declared the policy absolutely forfeited. It took a more generous course ; the policy was admitted, as a claim, but from the amount that would have come to the legatees, if all had been in order, the difference of premium between 35 and 42, for 42 years, with interest and compound interest thereon, from the period that each premium became due, was deducted. The legatees thus received not more than half the nominal amount stated on the policy.

11 Both ladies have since been married. "No cards." Were we a newspaper proprietor, we should charge two and sixpence additional for the two last words of this announcement.

seat of learning and refreshment, formed an important part of the population of the colony ; at least they are described as so being, both in an article of the Quarterly Review upon the London and North- Western Railway of 1849, and in Sir Frauds Head,s ', Stokers and Pokers," published in the identical same year, and almost in the identical same month. Nevertheless the former names seventy-five pigs and piglings as members of the refreshment establishment; but Sir Francis disputes the figures, raises it ten higher, and not only insists that they were eighty-five in number, but that each was converted, in his or her turn every year, into pork pies and sausage roly-polies. But whichever amount be the correct one, let the pork pies and sausage roly-polies rest in peace. The indigestions of which they were, in their day, the *tererinae causae*, have long since passed away. Let no rude attempt be made to re-produce them.

Thanks to increased lines of railway at Wolverton, both "main" and "siding," thanks also to signalling so improved in principle, and so minute in action — signalling which embraces the visible "arm," by day, the visible tri-coloured lamp at night, the audible "fog" and never-failing, ever truthful Electricity— express and other trains can, and do dart past the fifty-two mile post placed at the eastern extremity of the station with celerity and certainty, the same as at any other part of the system.

Francesca Marton. Attic and Area

This is not a primary source but an interesting footnote to the phenomenon of the Wolverton Refreshment Rooms. An early 20th century writer Margaret Bellasis (1884-1961), apparently a cousin of the much more famous Robert Graves, wrote a number of popular novels. This one, published in 1944, drew upon the historical information about the Refreshment Rooms to draw the central character of this novel, and certainly drew her material from Sir Francis Bond Head's description of the Wolverton Refreshment Rooms and the seven very young ladies who waited upon the passengers. The name Marton chose for one of the girls was Hannah Mary Christmas

For the historical record, the names of the "seven young ladies" working at the Refreshment Rooms in 1851 (not necessarily the same ones who were there in 1849) were:

Louisa Bryant	23
Janet Knight	26
Adelaide Halse	13
Eliza Gawcutt	16
Fanny Swan	17

Eliza Robertson   18
Annie Knight   17

This review of the book, written in 1948, comes from *The Tablet*.

Early Victorian London is a difficult territory for a novelist to make his own. Association is potent, and Attic and Area, crowded as it is with the raw material of almost any Dickens novel, begins with a liability. Miss Marton has done her work almost too well. Her industry has brilliantly reconstructed the world of crowded alleys and the new railways, of lawyers' chambers and basement kitchens, of fairs and theatre galleries, all providing the background of a year in service of a girl from Kent. And the characters of her novel, and all that happens to them, evoke, say, Great Expectations, with more than an occasional hint of Charlotte Bronte at work in the wings.

Such a first judgment is inevitable, but it is far from just to Attic and Area, which is a carefully devised and enthusiastically written novel, as welcome among its contemporaries as would be, in our dehydrated days, the solid meals it often describes. The blousy housekeeper and the disinherited son, the hypocritical preacher and the light-hearted medical student, the lovely lady of the manor and the slavey among the blackbeetles belowstairs: we have met them all before, and we are scarcely surprised at the missing letters, the unlooked for reconciliations or the double line drawn at the end. What is most remarkable about Miss Marton's novel is not so much its material as the use she makes of it. Examine the ingredients, certainly, but judge of their effect at the consumer's end. Attic and Area is rich, unrationed, done to a turn.

PART 10

THE CLOSE
OF THE DECADE

In this section

- from The Times, December 22nd 1849

Later in the century the railway company became active, even dominant, in the provision of amenities for Wolverton's population, but they were initially slow to respond. They did build a reading room and a market house quite early but pressure for schools and a church came from the Radcliffe Trust, as has been shown in the previous section. Pressure for a Mechanics Institute came from the workers themselves, most of whom, already literate, were hungry for more knowledge. At Christmas in 1849 a huge event was organised to promote the institute and is described in this lengthy Times report. Over 1000 people gathered in the erecting shop at Wolverton, which had been cleared of machinery for the occasion. It was a long evening of speeches but it did close the decade with a vision for the future of Wolverton. The illustration comes from the London Illustrated News. In the end, it took until 1864 for the Science and Art Institute to open its doors.

PART 10: THE CLOSE OF THE DECADE

Wolverton Mechanics Institute

THE WOLVERTON MECHANICS' INSTITUTE SOIRÉE.

The Times, Saturday, December 22nd 1849
(from our own reporter)

A *soirée* in aid of the Wolverton Mechanics' Institute was given yesterday at the works of the London and North Western Railway Company near the station. An event so interesting, not only to the workpeople, but to their employers, who have so kindly and strenuously assisted the establishment and permanent welfare of the institute, was rendered attractive to a wider circle by the report that several distinguished members of Parliament would be present and take part in the proceedings. If a numerous attendance on this occasion can give all the aid that is required, no doubt the Mechanics' Institute of Wolverton will be the most prosperous in the kingdom. The *soirée* was held in one of those great piles of buildings which have been erected by the company for the construction, repairing and cure of steam engines, in all their various stages from birth to old age. A well-proportioned and lofty shed of substantial bricks and mortar (130feet long by 90 feet broad) was selected from several larger apartments as the *salon*. It answered the purpose admirably, being lofty and capacious, and cool, despite the fuming heat of tea kettles and the crowd of guests who thronged it. Along the clean, whitewashed walls were ranged wreaths of holly, ivy, palm, and other evergreens, festooning the iron pipes and pillars. The light metal shafts which support the fragile looking but substantial roof were surrounded also by a quantity of the same simple decorations but here and there some grim wheel, with cogteeth, or eccentric bit of machinery that could not be got out of the way, thrust its spokes, or legs and arms, up through the leaves, and put one in mind of a savage lurking in ambush. A platform at one end raised the *élite* of the company into fair view of the audience, and afforded room for the musicians. The building was well and handsomely illuminated. Over the chair was a crown. And the Royal initials in gas, and there was no dearth of fiery devices along the walls – horns of plenty, stars, and wreaths, and various species of gas lamps shining through a liberal display of evergreen arches and festoons and union jacks in great variety. Teacups and saucers were laid for upwards of 14,000; and soon after 6 o'clock every place at the spacious tables was occupied. Men and women, all in their best, with smiling happy faces, thronged in, till upwards of 1,500 persons were collected in the building, exclusive of those who attended as guests. Admission was by tickets, which cost 6d. or 1s. each.

Mr. McConnell, superintendent of the locomotive department took the chair at 6 o'clock, being attended by Mr. G.C. Glyn, M.P., chairman of the company, Mr. Smith, Mr. Barrow, Mr. Lucy, Mayor of Birmingham, Mr. R. Creed, Mr. J.L. Prevost, and Mr. Earle, directors, Captain Huish, manager, Mr. Stewart, Secretary &c. After grace a tremendous clatter of cups and plates took place, and lasted with unabated vigour for half an hour. Gigantic teapots and cauldrons made their appearance from the steam furnace close at hand, and were emptied of their contents as fast as the attendants could bear them along the defiles of tables. The piles of bread and butter, biscuit and cake, were raised and destroyed, and raised again in quick succession, and all was good humour, enjoyment, and loud, but not boisterous contentment, till even the little children, of whom there was a rather large supply, as we suppose they could not be left at home, were forced to cry "Hold – enough."

Thanks having been returned by the Rev. Mr. Fremantle, of the new church built there by the company, the musical gentlemen, under the direction of Mr. Bruton, favoured the company with "Non nobis Domine", which was considerably approved of.

Mr. McConnell (the chairman) proceeded to observe, it was a most gratifying and cheering scene to look around and see the numerous and cordial friends assembled on that occasion. It was the first held to support a Railway Mechanics' Institution, and it was at Wolverton such a meeting should properly be held; for it was there the first town of railway servants had ever been established. It was there, too, a mechanics' institute might be expected to flourish, but he regretted that as compared to other places, they had not made the progress which might have been expected, and had not kept pace with similar institutions on other railways. But other times were coming, and the present attempt was proof of the spirit pervading the people at Wolverton. It might be said the town itself was altogether the offspring of the railway. There were employed there no less than 500 mechanics who were engaged on the work of 220 engines, running upwards of 3,000,000 miles in the course of the year, and conveying upwards of 1,000 tons per week. Such a working stock, it was evident, required a large amount of mechanical force to keep it in order and repair. Artisans of nearly every class were congregated together, and if Wolverton were transported tomorrow to the wilds of America there existed within it all the elements of production necessary for the comforts of life. (Cheers.) Everyone acquainted with its history would admit that its population had been most exemplary in conduct, and that, considering the many districts and parts of the kingdom from which they came, it was really gratifying to find how few causes of there were complaint against them. (Cheers.) The people and workmen had, indeed, been orderly, respectable, and well-conducted throughout. (Loud cheers.) It must not be forgotten that the liberality of the directors had placed the means of education with the reach of all of them and had afforded them an opportunity of attending divine service in the church built for the purpose. All the orderly character to which he referred was due to the very efficient service of the clergymen appointed to superintend their secular and religious education. (Hear, hear.) He rejoiced at such a meeting as the present; independently of the laudable object they had met to serve, the social repast they had just enjoyed enabled them to cultivate kindly feelings with their brother workmen (cheers), and to do away with those little jealousies which must exist in all great establishments. The institute had been in existence since 1840, but unfortunately it had not been successful, owing to want of a proper mode of action among the men; but they were now more united, and the example of the large mechanics' institutes in the manufacturing towns had had its effects. The great Exhibition of Manufactures would no doubt stimulate mechanics to use their native talent, but they could never put forth their powers till they were enabled to do so by mechanics' institutes. In conclusion, he might observe, that the London and North Western Company were among the first to encourage education among their working men, and the chairman was entitled to the highest praise for the uniform attention he had bestowed in increasing their comforts and enlightening their minds in every way he could by providing teachers and churches, and by seeing that teachers and churches fulfilled their ends. He had great pleasure in proposing for their consideration and applause "Prosperity to the London and North Western Railway Company, Chairman and Directors." (Loud and continuous cheering.)

Mr. Glyn rose to return thanks for the enthusiastic manner in which they had received the sentiment conveyed by the chairman. It had been his good fortune, on more than one occasion, to be present at Wolverton during these interesting celebrations. He had witnessed the opening of the schools, and had assisted at the dedication of their church; but on no occasion had he ever felt such heartfelt gratification as the present. On those former occasions he and his colleagues had attended to discharge, as trustees of the company, those duties and responsibilities which their situation imposed upon them. The company had thought it right, considering the mixed assemblage collected at Wolverton, that the schools should be opened on such a principle as would allow the admission of children of parents of all religious denominations. (Cheers.) They had also thought it right to meet the liberality of the Radcliffe Trustees, and take measures to support the church, and to extend those of their servants who were of the established church, and to the town and surrounding districts, the scriptural benefits to be derived from it. But he regarded the present ceremony with stronger feelings – and cold must be the heart who would not – because it appealed to the heart, and it conveyed to his mind corroborative testimony that those who were present appreciated the efforts of the company and that apart from those efforts had arisen that movement which they were now spontaneously carrying forward. (Hear, hear.) They were engaged in a great and noble work, but, although they were so engaged, he entreated them to reflect that after the provision of proper spiritual instruction, their highest duty as citizens and parents, was not the further education of themselves, but of their children. (Hear, hear.) Education was the groundwork of everything valuable in after life. Let them conceive what a basis it lay down for the rising generation. Let them remember that they were living in a country where the lowest among them might, if properly educated, arise in the race of life to the highest rank – that in this happy country – blessed be to God for it! – no degree, no grade, no exclusion existed, which prevented the well-educated youth from taking up a high position, such as his father, not so well-educated, could never have arrived at. (Cheers.) He needed not to remind them of the instances of the truth of that assertion, but there was one whom he could not refrain from mentioning here, because it had been his lot to have been thrown much into contact with him – he alluded to the late George Stephenson. (Cheers.) He had his failings, which of them had not (cheers)? But he (Mr. Glyn) held that the rise, the life, and the position of that man had been an honour to himself and to the country to which he belonged. He had heard him often detail passages in his interesting life – the privation which he had endured, and the industry with which he struggled against the anxiety of his early commencement; but what had struck him (Mr. Glyn) most, and had made the deepest impression on him, was when he recounted the zeal, the toil, and the privations he underwent to ensure the best education for his only son. (Cheers.) He appealed to them that if he (Mr. Stephenson) had not been repaid for that toil and for those privations? Had not that son repaid everything a father could have done? And did he not know bear a European character and estimation? (Cheers.) Let them rely on it that the cost of education would be one which they would never regret to have paid. (Cheers.) But there was another cause of congratulation which he could not pass over. They had there, to celebrate the occasion on which they had met, gentlemen who would do honour to any assembly, and he confessed that, having with them those not so immediately connected with the railways as themselves, he was anxious to occupy a little of their time, and to bespeak their kinder consideration for those who held in their hands the administration of those great undertakings. They were assailed with cries on every side; and far be it for him in any assemblage to extenuate or deny that any deserved reproaches had been cast on those who ought, in the position they held, to have considered themselves the trustees and representatives of others, and not the mere promoters of their own selfish end. (Cheers.) But, while that was so, was it right or fair that all who had from the earliest date of railway enterprise had striven to mature the system and bring it to the present point of perfection should be mixed up in one indiscriminate torrent of abuse? (Cheers.) Notwithstanding all that had been done – that towns had been erected where hamlets had not existed before, and that arrangements had been made which enabled a gentleman to step from his carriage at Euston square, and to travel from one end of England to the other, to proceed from London to York, or Montrose, should they all be heaped up together in one torrent of abuse and be excluded from a fair participation in the encomiums which, in his opinion, they deserved? Hear.) But in all these arrangements they had never asked for any assistance from Government. The railways never had had the slightest assistance from Government.(Hear, hear.) Government had only thought of taxing them. They never had to thank Parliament for the slightest aid – Parliament had only interfered to diminish their rates and tolls. (Cheers.) But he had – and he rejoiced to have an opportunity of saying it publicly to thank the gentlemen assembled for the consideration they had ever given their employers, and that in every proceeding they had commenced for the improvement of that understanding, from first to last, they had received that untiring co-operation of their servants, and whether he looked to those at Wolverton or to those whose avocations prevented their presence that night – the guards and drivers – he had to declare that the company had received the most unflinching co-operation, and that through the means of their servants their present system had been laid down – a system which Government interferences might mar, but which Government interference could not improve. (Loud cheers.) He could not

224

sit down without doing justice to the Rev. Mr. Waite (sic), and expressing publicly the satisfaction he felt, and the thanks the company conceived due to him, for the way in which he had carried out the wishes of the directors. Although a minister of the Church of England, he had not hesitated, in his administration of the affairs of the schools, to open them to children of all denominations. (Hear, hear.) He (Mr. Glyn) had now presided over the London and North Western Company for many years. He knew not the course of events, or what might be coming to touch and affect railway interests, but whatever that course might be, or whatever might befall them, he should always feel it an honour to be connected to a company of which the employers and employees could meet in the way they had done that evening. (Tremendous cheering.)

Mr. Barron, in a few words, proposed the speedy operation of the Buckingham Railway, in connexion with the health of the county member.

Sir H. Verney acknowledged the compliment, and having expressed the gratification he felt at being present on so agreeable occasion, impressed on the audience the paramount importance of the holy Scriptures as the source of all real knowledge. (Loud cheers.)

Mr. Lucy (Mayor of Birmingham) proposed very briefly, "The Working Staff of the London and North Western Railway." (Cheers.)

Captain Huish returned thanks, and in doing so enlarged on the varied and extensive duties of the department he superintended, and on the immense interests committed to the charge of the company. There were some present who might not be aware of the magnitude of the undertaking. The company employed rather more than 10,000 persons, and about 140,000 people travelled their line every week. Now, the public were not, generally speaking, very grateful. Every one of these people, on an average, had three parcels of some kind or another, or, in other words, there were about half a million of bandboxes, and carpet bags, and such articles conveyed by the line every week; and when it was considered that of their passengers a large proportion were ladies, who almost invariably left everything behind them (laughter), and when he told them that the board of the company had not to pay for one of those parcels oftener than once in three months, they might be deeply – they ought to be deeply – grateful to their 10,000 servants. From the commencement of the railway, 100,000,000 persons had travelled on it, and, with the exception of one melancholy event near that spot, he would ask them, could any conceivable invention of man have produced a greater amount of safety? (Cheers.) There were 900 policemen on the line, and the least neglect of duty of any one of them might cause the most fatal accident, and yet the amount of loss of life was almost inconceivably small. Having alluded to the practical lessons in order and regularity taught to the people by railways, he proceeded to urge on his audience the necessity of avoiding agitators and evil counsellors, and regarding the Bible as the sole study by which their advance in secular knowledge could be made peaceful or useful, and concluded by introducing to the meeting "Mr. George Cruikshank, the Hogarth of the 19th. Century."

Mr. Cruikshank, who was received with lopud applause, returned thanks for himself and the guests of the evening. As a working man himself he was glad to be present on such an occasion. He had worked hard himself, and he thanked God for it, and that he had been able to do so. The directors, he was sure, wished them well. They would give their workmen their due. (Cheers.) If anything would ever raise England it would be the cheap system. (Loud cheering.) "A fair day's wages for a fair day's work." – that was the motto. (Great applause.) As an artist he could assure them he never saw a more beautiful picture than the present; he saw not only the front he saw the back of the canvass; and more especially glad he was to see so many women present, for they might be certain, that though it had been said that women were at the bottom of every mischief, there never yet was any great social movement in which a woman had not taken part. (Cheers.)

The Rev. Mr. Fremantle, in speaking to the same toast, vindicated the character of railway labourers, and declared there were no men he would sooner have to deal with.

After a few words from Dr. MacKay, who was introduced to the meeting by Mr. Cruikshank.

Captain Huish proposed the health of "The Press". They might be of opinion that the press just now bore rather hard on the railway interest. When they were hard set for a leader, and Parliament was not sitting, they set to work to abuse the railways. (Cheers and laughter.) But still the daily press had done them good service. True, it was often made the means of intimidation. For instance, if any of the ladies, of whom he had spoken

226

before, did not find her bandbox or bag forthcoming, she wrote to hi at once – "Sir, if my box is not returned in two days I'll write to The Times.' 9loud laughter.) That was the panacea for all their evils. (Renewed laughter.) Some time ago a gentleman was smuggling a suckling pig in one of the carriages. One of the porters saw it, and said he must pay 6d. for it. "What! Am I to pay for a sucking pig, when you let children in arms go free?" (Laughter.) And if the money was not returned he supposed the next letter would be "I'll write to The Times." (Cheers and laughter.)

Mr. Watkin, Assistant Secretary, and Mr. Henderson, one of the workmen, addressed the meeting, and other gentlemen were preparing to get on their legs, when we were obliged to get on ours to catch the last train; but their audience had greatly diminished, as the speeches were long, and could not be heard in the remoter parts of the hall, and a dance and supper elsewhere had powerful attractions. All the arrangements, which were under the management of Mrs. Hibbert, were very creditable to her taste and industry. A long programme of music was still undisposed of at half-past ten o'clock.

Lightning Source UK Ltd.
Milton Keynes UK
UKOW04f0708141214

243063UK00002B/138/P